Chemotherapy

International Journal of Experimental and Clinical Chemotherapy

Founded 1959 as 'Chemotherapia'

KARGER

Printed in Switzerland
on acid-free paper by
Thür AG Offsetdruck, Pratteln

Appears bimonthly:
1 volume per year (6 issues)

Submission: Only original papers written in English are considered and should be sent to:

S. Karger AG
Editorial Dept. 'Chemotherapy'
P.O. Box
CH–4009 Basel (Switzerland)

Manuscripts should be submitted in duplicate (with two sets of illustrations of which one is an original), typewritten double-spaced on one side of the paper, with a wide margin.

Conditions: All manuscripts are subject to editorial review. Manuscripts are received with the explicit understanding that they are not under simultaneous consideration by any other publication. Submission of an article for publication implies the transfer of the copyright from the author to the publisher upon acceptance. Accepted papers become the permanent property of 'Chemotherapy' and may not be reproduced by any means, in whole or in part, without the written consent of the publisher. It is the author's responsibility to obtain permission to reproduce illustrations, tables, etc. from other publications.

Arrangement
Title page: The first page of each paper should indicate the title (main title underlined), the authors' names, and the institute where the work was conducted. A short title for use as running head as well as the full address of the author to whom correspondence should be sent are also required.
Full address: The exact postal address complete with postal code must be given at the bottom of the title page. Please also supply Phone, Fax, and E-mail numbers.
Key words: For indexing purposes, a list of 3–10 key words in English is essential.
Abstract: Each paper needs an abstract of up to 10 lines.
Small type: Paragraphs which can or must be set in smaller type (case histories, test methods, etc.) should be indicated with a 'p' (petit) in the margin on the left-hand side.
Footnotes: Avoid footnotes. When essential, they are numbered consecutively and typed at the foot of the appropriate page.
Tables and illustrations: Tables and illustrations (both numbered in Arabic numerals) should be prepared on separate sheets. Tables require a heading and figures a legend, also prepared on a separate sheet. For the reproduction of illustrations, only good drawings and original photographs can be accepted; negatives or photocopies cannot be used. Due to technical reasons, figures with a screen background should not be submitted. When possible, group several illustrations on one block for reproduction (max. size 144 × 196 mm) or provide crop marks. On the back of each illustration, indicate its number, the author's name, and 'top' with a soft pencil.

Color illustrations: Up to 6 color illustrations per page can now be integrated with the text at the reduced price of CHF 600.–/ USD 545.00 per page. Color illustrations are reproduced at the author's expense.

References: In the text identify references by Arabic numerals [in square brackets]. Material submitted for publication but not yet accepted should be noted as 'unpublished data' and not be included in the reference list. The list of references should include only those publications which are cited in the text. Do not alphabetize; number references in the order in which they are first mentioned in the text. The surnames of the authors followed by initials should be given. There should be no punctuation other than a comma to separate the authors. Cite all authors, 'et al' is not sufficient. Abbreviate journal names according to the Index Medicus system. (Also see International Committee of Medical Journal Editors: Uniform requirements for manuscripts submitted to biomedical journals. Br Med J 1988;296:401–405.)
Examples
(a) Papers published in periodicals: Kauffman HF, van der Heide S, Beaumont F, Blok H, de Vries K: Class-specific antibody determination against *Aspergillus fumigatus* by means of the enzyme-linked immunosorbent assay. III. Comparative study: IgG, IgA, IgM ELISA titers, precipitating antibodies and IgE binding after fractionation of the antigen. Int Arch Allergy Appl Immunol 1986;80:300–306.
(b) Monographs: Matthews DE, Farewell VT: Using and Understanding Medical Statistics. Basel, Karger, 1985.
(c) Edited books: Hardy WD Jr, Essex M: FeLV-induced feline acquired immune deficiency syndrome: A model for human AIDS; in Klein E (ed): Acquired Immunodeficiency Syndrome. Prog Allergy. Basel, Karger, 1986, vol 37, pp 353–376.
Further details: Authors may consult our leaflet 'Rules for the Preparation of Manuscripts', available on request from the Publisher, S. Karger AG, P.O. Box, CH–4009 Basel (Switzerland).

Page charges: There is no page charge for papers of 6 or fewer printed pages (including tables, illustrations and references). Each additional complete or partial page is charged to the author at CHF 245.– / USD 196.00. The allotted size of a paper is equal to approx. 12 manuscript pages (including tables, illustrations and references).

Galley proofs: Unless indicated otherwise, galley proofs are sent to the first-named author and should be returned with the least possible delay. Alterations made in galley proofs, other than the correction of printer's errors, are charged to the author. No page proofs are supplied.

Reprints: Order forms and a price list are sent with the galley proofs. Orders submitted after the issue is printed are subject to considerably higher prices. Allow five weeks from date of publication for delivery of reprints.

KARGER

Publication data: 'Chemotherapy' is published 6 times annually. Volume 42 with 6 issues appears in 1996.

Subscription rates: Subscriptions run for a full calendar year. Prices are given per volume.

Personal subscription:

Subscription	+ *postage and handling*	
CHF 330.–	+ CHF 16.60	Europe
CHF 330.–	+ CHF 25.50	Overseas
USD 253.80	+ USD 16.60	

(Must be in the name of, billed to, and paid by an individual. Order must be marked 'personal subscription'.)

Institutional subscription:

Subscription	+ *postage and handling*	
CHF 550.–	+ CHF 27.60	Europe
CHF 550.–	+ CHF 42.–	Overseas
USD 423.00	+ USD 27.60	

Airmail postage: CHF 36.–, USD 30.00. (Extra per volume.)

Microforms: Please direct all enquiries and orders to:
University Microfilms Inc.
300 North Zeeb Road, P.O. Box 1346
Ann Arbor, MI 48106-1346 (USA)

Advertising: Correspondence and rate requests should be addressed to the Publisher.

Subscription orders: Orders can be placed at agencies, bookstores, directly with the Publisher S. Karger AG, P.O. Box, CH – 4009 Basel (Switzerland) or with any of the following distributors:

Deutschland: S. Karger GmbH, Postfach
D – 79095 Freiburg

France: Librairie Luginbühl
36, bd de Latour-Maubourg
F–75007 Paris

Great Britain, Ireland: S. Karger Publishers
11 Ravenscroft Road
Beckenham, Kent BR3 4TP

USA: S. Karger Publishers, Inc.
26 West Avon Road, P.O. Box 529
Farmington, CT 06085

India, Bangladesh, Sri Lanka:
S. Karger India Subscription Agency
B – 5/132 Safdarjung Enclave
New Delhi-110029

Japan: Katakura Libri, Inc.
Take 1 Bldg, 2nd Floor
4-5-7 Yushima, Bunkyo-ku
Tokyo 113

Change of address: Both old and new address should be sent to the subscription source.

Single issues and back volumes: Information can be obtained through the Publisher.

Bibliographic indices: This journal is regularly listed in bibliographic services, including *Current Contents®*.

Book Reviews: Books are accepted for review by special agreement.

KARGER

A comprehensive update on the Th1/Th2 dichotomy

Th1 and Th2 Cells in Health and Disease

Editor: Sergio Romagnani

Th1 and Th2 cells, representing extremely polarized modalities of the specific effector immune response, play an important role in both protection against exogenous insults and the induction of several immunopathological reactions. This book provides an authoritative update on the developments in this fast-moving field from both basic and clinical perspectives. In 12 detailed chapters, leading immunologists highlight the physiological meaning of Th1 and Th2 cells, their functional properties and surface markers, the mechanisms responsible for their development, as well as their roles in protection against bacteria, protozoans, fungi, viruses and helminths. Also discussed are the roles of Th1 and Th2 cells in some immunopathological conditions, such as autoimmune and neurological disorders, allergy and HIV infection. By combining for the first time complete information on basic mechanisms together with the clinical implications of Th1/Th2-dominated immune responses, this book will be appreciated by a broad spectrum of readers, including immunobiologists and clinicians from many medical fields.

Contents

Fields of Interest: Immunology; Infectious Diseases, Cell Biology, Molecular Biology, AIDS, Allergy, Neurology, Microbiology, Parasitology

Chemical Immunology, Vol. 63
Series Editors: Adorini, L. (Milan); Arai, K. (Tokyo); Berek, C. (Berlin); Capra, J.D. (Dallas, Tex.); Schmitt-Verhulst, A.-M. (Marseille); Waksman, B.H. (New York, N.Y.)
ISSN 1015–0145

Th1 and Th2 Cells in Health and Disease
Editor: Romagnani, S. (Florence)
XII + 222 p., 17 fig., 9 tab., hard cover, 1996
Institutions: CHF 258.– / DEM 309.– / USD 224.50
Individuals: CHF 129.– / DEM 154.– / USD 112.25
Prices subject to change
DEM price for Germany, USD price for USA only
ISBN 3–8055–6241–1

KI 96233

KARGER

Quinolones in Everyday Clinical Practice

7th European Congress of Clinical Microbiology and Infectious Diseases (ECCMID)

Satellite Symposium,
March 27, 1995,
Vienna, Austria

Guest Editor
W. Graninger, Vienna

11 figures, 23 tables, 1996

KARGER

S. Karger
Medical and Scientific Publishers
Basel · Freiburg · Paris · London
New York · New Delhi · Bangkok
Singapore · Tokyo · Sydney

Drug Dosage
The authors and the publisher have exerted every effort to ensure that drug selection and dosage set forth in this text are in accord with current recommendations and practice at the time of publication. However, in view of ongoing research, changes in government regulations, and the constant flow of information relating to drug therapy and drug reactions, the reader is urged to check the package insert for each drug for any change in indications and dosage and for added warnings and precautions. This is particularly important when the recommended agent is a new and/or infrequently employed drug.

Printed in Switzerland on acid-free paper by
Thür AG Offsetdruck, Pratteln
ISBN 3-8055-6299-3

Chemotherapy

Vol. 42, Supplement 1, 1996

Contents

KARGER

E-Mail karger@karger.ch
Fax + 41 61 306 12 34

Chemotherapy

Chemotherapy 1996;42(suppl 1):1–9

Fleroxacin Overview

K.G. Naber

Department of Urology,
University of Munich Medical
School, Elisabeth Hospital,
Straubing (Germany)

Key Words
Fleroxacin
Quinolone
Efficacy
Tolerability

Abstract
Fleroxacin is a new oral and intravenous trifluorinated 4-quinolone, which acts by inhibiting the essential bacterial enzyme DNA gyrase. Fleroxacin exhibits a broad spectrum of action, characterized by pronounced activity against aerobic gram-negative bacteria, but also against gram-positive pathogens such as staphylococci. Fleroxacin is distinguished by its excellent bioavailability, high concentrations in the plasma and other body fluids, good tissue penetration, and a long half-life of 10–12 h, thus allowing once-a-day administration. A single oral dose of 400 mg fleroxacin is effective in uncomplicated cystitis in women, uncomplicated gonococcal infections, bacterial enteritis, and traveler's diarrhea. A single daily dose of 200 mg administered for 3 days is effective in uncomplicated urinary tract infection (UTI), while longer treatment and higher doses may be required in acute uncomplicated pyelonephritis and complicated UTI. Skin, soft tissue, bone and joint infections, and lower respiratory tract infections including exacerbation of chronic bronchitis and non-pneumococcal pneumonia are further indications for fleroxacin.

This is a brief update on fleroxacin, a new oral and intravenous trifluorinated 4-quinolone, which acts by inhibiting the essential bacterial enzyme DNA gyrase.

Microbiology

Fleroxacin exhibits a broad spectrum of action, characterized by pronounced activity against aerobic gram-negative bacteria.

KARGER

E-Mail karger@karger.ch
Fax + 41 61 306 12 34

0009–3157/96/0427–0001$10.00/0

K.G. Naber, MD
Urologic Clinic, Elisabeth Hospital, Straubing
University of Munich Medical School
Teaching Hospital
D–94315 Straubing (Germany)

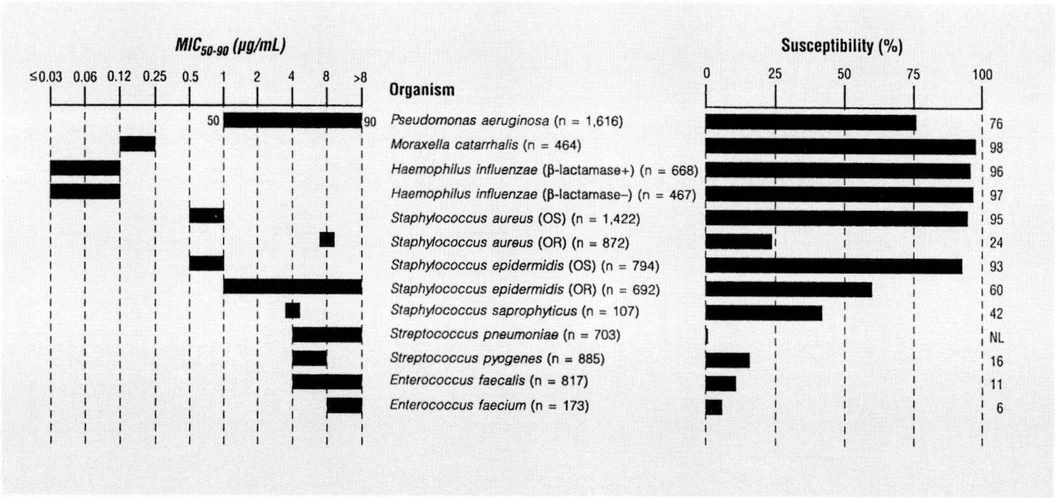

Fig. 1. In vitro activity of fleroxacin against clinical bacterial isolates of gram-negative and gram-positive species [adapted from ref. 1]. The following susceptibility MIC breakpoint was used: ≤2 µg/ml according to NCCLS [27]. OS = Oxacillin susceptible; OR = oxacillin resistant.

Gram-negative bacteria like *Neisseria gonorrhoeae, Aeromonas hydrophila,* and *Haemophilus ducreyi* are highly sensitive to fleroxacin and, accordingly, the MIC_{90} (minimal concentration that inhibits the growth of 90% of a specific strain) for these pathogens is very low (fig. 1). Another sensitive group are the Enterobacteriaceae. Although Enterobacteriaceae resistant to quinolones have been detected in some countries, in most the MIC_{90} is still very low (fig. 2). The third group of bacteria sensitive to fleroxacin include certain gram-positive bacteria like *Staphylococcus epidermidis* and *Staphylococcus aureus,* both the methicillin-resistant and the methicillin-sensitive strains (fig. 1). Fleroxacin is less active against *Streptococcus pneumoniae.*

According to our own experience, uropathogens which cause complicated urinary tract infections (UTIs), including Enterobacteriaceae, *Escherichia coli, Enterobacter, Klebsiella, Proteus* and other gram-negative microorganisms, are all sensitive to fleroxacin, with MICs of <1–2 µg/ml. A comparison of the MICs of fleroxacin and ofloxacin indicated that susceptibility to these two drugs appears to be parallel. This is also true for ciprofloxacin and pefloxacin.

A recently published multicenter comparative in vitro study of fluoroquinolones against 25,129 gram-positive and gram-negative clinical isolates confirmed the broad spectrum of activity of fleroxacin and other fluoroquinolones [1].

In general, the in vitro activity of fleroxacin is similar to that of other quinolones such as ofloxacin, norfloxacin, pefloxacin, enoxacin, and lomefloxacin. The vast majority of key microorganisms is susceptible to fleroxacin, and this fact, coupled with the excellent tissue penetration and high serum and urine levels achieved, results in its excellent clinical and bacteriological efficacy.

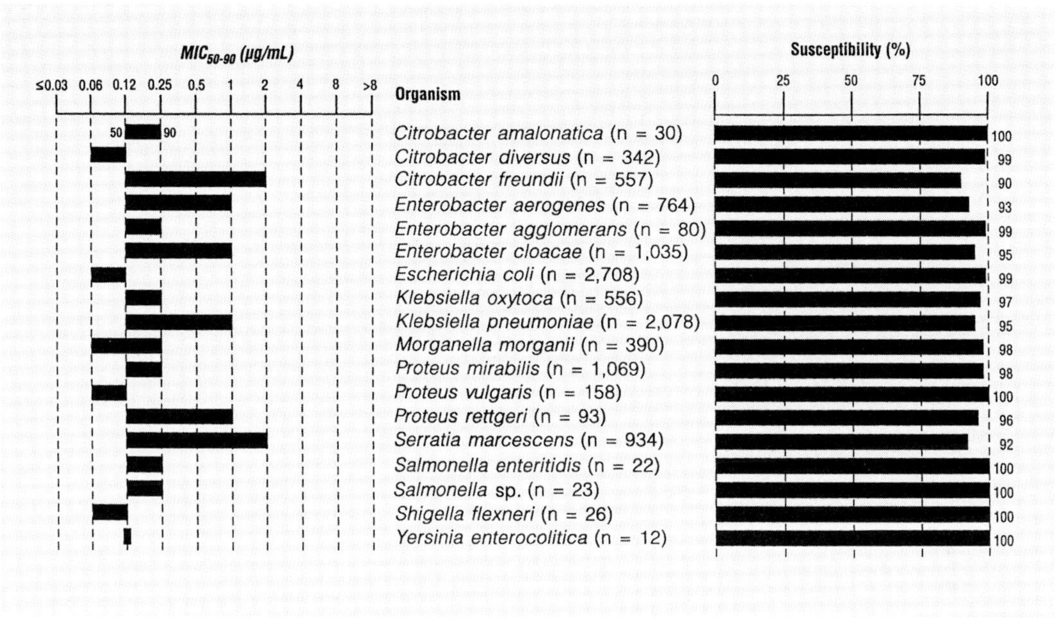

Fig. 2. In vitro activity of fleroxacin against clinical bacterial isolates of Enterobacteriaceae [adapted from ref. 1]. The following susceptibility MIC breakpoint was used: ≤2 µg/ml according to NCCLS [27].

Pharmacokinetic Properties

The standard dosage of fleroxacin is 200 or 400 mg once a day orally, or 400 mg once a day administered intravenously within 1 h. The intravenous infusion of 400 mg over a 1-hour period produces a mean peak concentration of 5–6 mg/l [2]. Plasma concentrations decrease slowly, since the elimination half-life is 12 h. Following oral administration of a 400-mg tablet, the plasma concentration profile is almost identical to that following intravenous dosing (fig. 3). This indicates that the bioavailability of fleroxacin is virtually 100%. Therefore, it is easy to switch from intravenous to oral administration, no dosage adjustments being necessary.

Compared to other quinolones, the bioavailability of fleroxacin (100%) is higher than that of ciprofloxacin (65–85%) and similar to that of pefloxacin (100%) and ofloxacin (95–100%). The mean peak concentration following the administration of 400 mg fleroxacin (5–6 mg/l) is higher than that of 400 mg norfloxacin (1.5 mg/l), 500 mg ciprofloxacin (2.3 mg/l), or 200 mg ofloxacin (3.6 mg/l). The elimination half-life of fleroxacin is about 12 h, the same as pefloxacin. Sparfloxacin has a half-life of 15–20 h, whereas most other quinolones like ciprofloxacin, norfloxacin, and ofloxacin have shorter half-lives between 3 and 7 h.

Quinolones are characterized by good tissue penetration, superior to that of β-lactam antibiotics or aminoglycosides. For example, 8 h after the administration of 400 mg fleroxacin, its concentration in lung tissue can reach levels as high as 8.9 mg/kg, with a tissue/plas-

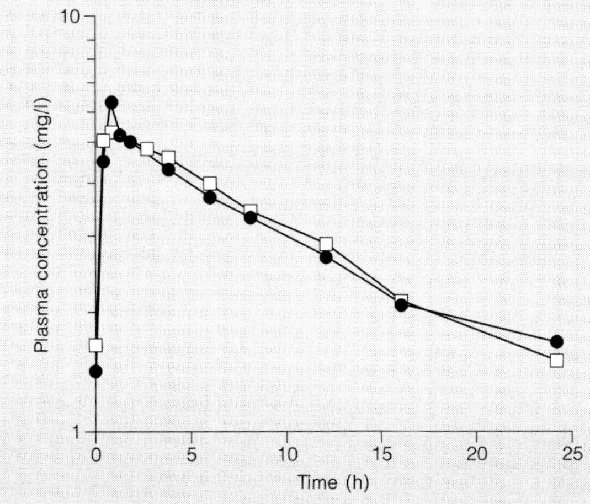

Fig. 3. Plasma concentration profile of fleroxacin 400 mg after oral (□) and intravenous (●) administration in healthy volunteers [from *Clin Drug Invest* 1995;9 (suppl 1):1–7, with permission].

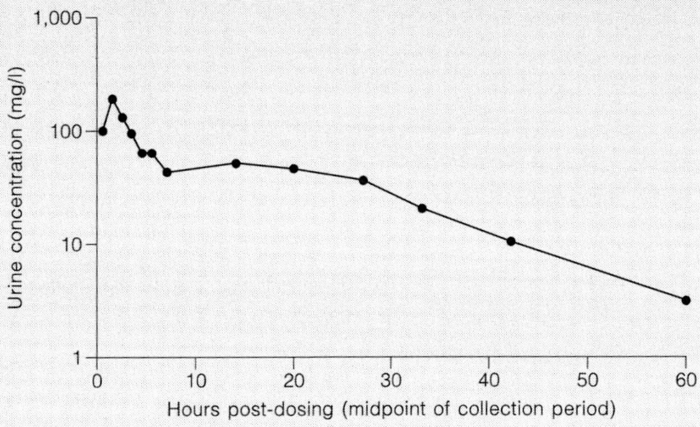

Fig. 4. Mean urine concentrations of fleroxacin following a single oral dose of 200 mg [adapted from ref. 7].

ma ratio of 2 [3]. This is due to the high peak concentration of fleroxacin in plasma, its long elimination half-life and good lung tissue penetration. Ciprofloxacin is also concentrated in lung tissue, but due to its lower plasma concentration and shorter half-life, the tissue concentration following oral administration of 500 mg ciprofloxacin is lower, i.e. about 5 mg/kg at 3 h [4]. Although the results of different methods of measurement may vary considerably, bone concentrations of fleroxacin remain at 2–4 mg/kg for up to 24 h [5].

Quinolones penetrate well into the genitourinary tract. The high concentrations achieved in the following organs after administration of 400 mg fleroxacin are similar to those found in the plasma: myometrium, 6.0 mg/kg; ovary, 6.4 mg/kg; seminal fluid after 12 h, 5.8 mg/l; in the fallopian tube, concentrations were somewhat lower (4.2 mg/kg). Concentrations in prostatic fluid 4 h after administration were lower (1.1 mg/l), although the penetration ratio is still higher (prostatic fluid/plasma ratio of approximately 0.25 to

Table 1. Interactions of quinolones with food and drugs [8]

	Effect of drug/food on pharmacokinetics of quinolone					Effect of quinolone on pharmacokinetics of drug (decrease of clearance, %)		
	decrease of bioavailability, %		decrease of AUC	increase of AUC	increase of t_{max}			
	antacids	sucralfate	% ranitidine	% cimetidine	% food	caffeine	theophylline	warfarin
Fleroxacin	25	24	no effect	25	no effect	no effect	<7	no effect
Ciprofloxacin	90	96	no effect	no effect	<10	40	20–30	
Enoxacin	85	88	40	20	<10	50–80	45–75	32
Lomefloxacin	20–50		no effect		70	<10	<8	no effect
Norfloxacin	98	98			<10	no effect	4–17	
Ofloxacin	50		no effect		75	no effect	4–12	
Pefloxacin	50			40	no effect		20–30	

0.5) than with β-lactam antibiotics, and fleroxacin levels are considerably higher in prostatic adenoma tissue (5.4 mg/kg).

In polymorphonuclear leukocytes, fleroxacin achieves intracellular concentrations that are approximately double the extracellular concentrations [6].

Following a single oral dose of 200 mg fleroxacin, urine concentrations exceeding the MIC_{90} for the major urinary pathogens are maintained for over 24 h (fig. 4) [7]. Determinations of the urinary bactericidal titer indicate that even after 5 days, active concentrations can still be found in the urine.

Interactions

Interactions of quinolones with food and other drugs vary and therefore each quinolone must be considered separately. A summary is provided in table 1 [8]. Fleroxacin does not affect caffeine or warfarin clearance, and decreases the clearance of theophylline by less than 7%. Antacids or sucralfate decrease the bioavailability of fleroxacin by 25%, and cimetidine increases the AUC of fleroxacin by 25%. Food, and Fe^{2+} or ranitidine have no effect on the t_{max} or AUC, respectively. In contrast, ciprofloxacin decreases the clearance of caffeine by 40%. Antacids and sucralfate decrease the bioavailability of ciprofloxacin by 90%, whereas ranitidine or cimetidine do not affect the AUC of ciprofloxacin. The interactions of enoxacin are to some extent more pronounced than those of ciprofloxacin. The interactions of other quinolones are less important.

Therapeutic Efficacy

UTIs

Fleroxacin is very effective in UTIs. Pooled data of studies in uncomplicated UTI show similar results with fleroxacin at a dosage of 200 mg for 3 or 7 days, or with a single dose of 400 mg: clinical cure rates of between 92 and 96% were obtained with bacteriological eradication rates of between 92 and 98% (fig. 5). In double-blind controlled studies, the bacteriological and clinical efficacy of fleroxacin (400 mg single dose or 200 mg for 3 or 7 days) was similar to that of ciprofloxacin 250 mg twice a day for 7 days [9, 10]. Single oral dose treatment of uncomplicated UTI in women of reproductive age is reviewed by Bailey [this volume].

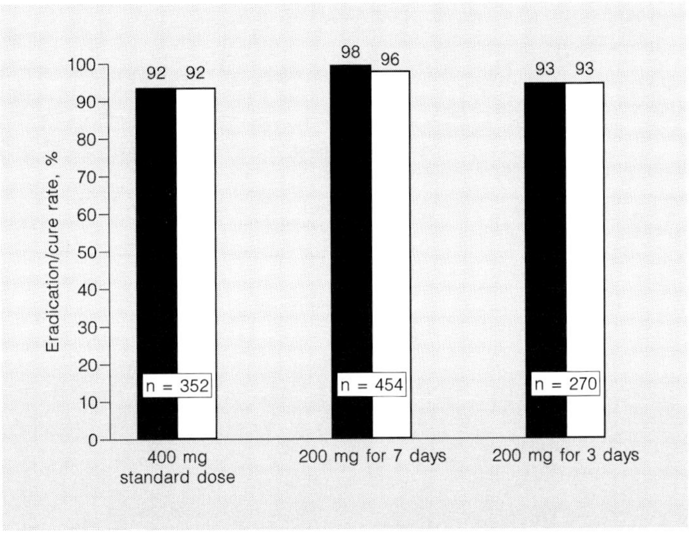

Fig. 5. Bacterial eradication (■) and clinical cure (□) rates in patients with uncomplicated UTI treated with a single dose of 400 mg fleroxacin or 200 mg once daily for 3 or 7 days. Pooled data from several studies.

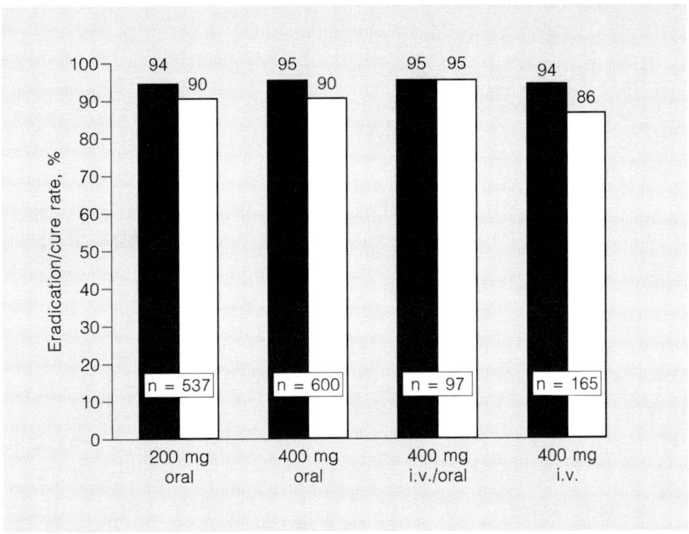

Fig. 6. Bacteriological eradication (■) and clinical cure (□) rates in patients with complicated UTI treated with 200 mg or 400 mg fleroxacin once daily by oral or intravenous (i.v.) routes. Pooled data from several studies.

In complicated UTI, good clinical results were obtained with the customary treatment duration of 7–14 days or longer. Pooled fleroxacin data from all studies show that in patients receiving 200 or 400 mg orally or intravenously once a day, clinical cure rates varied between 86 and 95%, and bacteriological eradication rates between 94 and 95% (fig. 6). The efficacy of fleroxacin in the treatment of complicated UTI is extensively reviewed by Giamarellou [this volume]. Results of individual studies demonstrated that clinical cure rates and bacteriological eradication rates with fleroxacin were comparable to those attained with norfloxacin 400 mg b.i.d. [11] or ciprofloxacin 500 mg b.i.d. [12].

Respiratory Tract Infections

In acute non-pneumococcal pneumonia, fleroxacin administered at an intravenous dose of 400 mg/day to 69 patients achieved bacteriological and clinical cure rates of 85 and 88%, respectively. The response was the same as that with ceftazidime at a dose of 0.5–2.0 mg/day [13]. Preliminary results of a new study of pneumonia are reported by Marklein [this volume]. This ongoing study encompasses 93 patients treated with fleroxacin (400 mg once a day) versus ciprofloxacin (400–500 mg b.i.d.). Clinical cure and improvement rates reached 73% with fleroxacin and 79% with ciprofloxacin.

In acute exacerbation of chronic bronchitis, clinical cure rates with fleroxacin at an oral dose of 400 mg/day administered for 7 days (n = 170) were higher than with amoxicillin at a dose of 500 mg three times a day (n = 168): 89 versus 80% [14, 15]. The differences in the overall bacteriological eradication rates showed that fleroxacin was clearly superior to amoxicillin (96 vs. 79%).

Skin and Soft-Tissue Infections

In various infections of the skin and soft tissue, i.e., skin abscesses, cellulitis, wound infections, skin and decubitus ulcers, or impetigo treated with 400 mg fleroxacin once a day for up to 3 weeks, clinical cure rates and overall bacteriological eradication rates of about 80% were achieved. These results were comparable to those reported for amoxicillin/clavulanic acid [16, 17] or ceftazidime [18].

Gastrointestinal Infections

The treatment of gastrointestinal infections with fleroxacin is reviewed in detail by Graninger et al. [this volume]. In bacterial enteritis, a single dose of 400 mg fleroxacin (n = 110) and a 3-day treatment (n = 112) achieved cure rates of 94 and 93%, respectively [19]. The use of placebo also led to respectable results in this indication (62%, n = 110), in traveler's diarrhea, a 1- or 2-day treatment with 400 mg fleroxacin orally was effective in 80% of the cases, and thus clearly superior to placebo, with 47% of the patients cured [20]. In typhoid fever, 1 week of treatment with 400 mg fleroxacin once a day led to results similar to the standard 2-week therapy with chloramphenicol; with a 2-week treatment with fleroxacin, cure rates of 100% were achieved [21]. Ongoing studies are investigating whether the treatment duration can be further reduced. In cholera infections, a single intravenous dose of fleroxacin cured 100% of the patients and was as effective as 500 mg tetracycline administered four times a day for 3 days [22].

Other Infections

Sexually transmitted diseases are still a severe problem all over the world. *N. gonorrhoeae* and *Chlamydia trachomatis* are among the most widespread infectious pathogens, occurring frequently in combination in the same patient. A single oral dose of 400 mg fleroxacin is highly effective in the treatment of uncomplicated gonorrhea and represents an oral alternative to ceftriaxone [23].

The efficacy of oral fleroxacin administered once a day was also evaluated in patients with either septic arthritis or acute or chronic osteomyelitis [24, 25], with a clinical cure rate of 62% (n = 39) and a bacteriological eradication rate of 79% (n = 39).

Fleroxacin is not only useful in the treatment of infections, but also in surgical prophylaxis. Gasser [this volume] has demonstrated that a single dose or short-time prophylaxis can be beneficial in transurethral surgery.

Safety

More than 8,000 patients have been treated with 200- and 400-mg oral doses of fleroxacin: 22% of these experienced at least one adverse event, possibly or probably related to the drug [26]. The side effect profile is similar to that seen with other fluoroquinolones: 11% of the adverse events involve the central nervous system (insomnia 6%, headache 3% and, less frequently, dizziness, lightheadedness, and nervousness); 11% affect the digestive system (the most frequent reaction being nausea in 7%) and 3% the skin (the most frequent side effect being photosensitivity/sunburn in 1%). Therefore, as with other quinolones, exposure to sunlight and artificial UV light should be avoided. Fleroxacin, like other quinolones, should also be used with caution in patients with known or suspected neurological disorders, since it may cause central nervous system stimulation, and on very rare occasions, seizures have occurred in such patients. The tolerability profile of intravenous fleroxacin 400 mg appears to be about the same as that of the same dose administered orally. It should be noted that the incidence of adverse events with fleroxacin has been observed to be dose-dependent.

Summary and Conclusion

Fleroxacin exhibits a broad spectrum of action, characterized by pronounced activity against aerobic gram-negative bacteria as well as against gram-positive pathogens such as staphylococci. The pharmacokinetics of oral fleroxacin are reflected in its excellent bioavailability, high concentrations in the plasma and other body fluids, good tissue penetration, and a long half-life of 10–12 h, which permits a once-a-day administration schedule. A single oral dose of 400 mg fleroxacin is effective in uncomplicated cystitis in women, uncomplicated gonococcal infections, bacterial enteritis, and traveler's diarrhea. A 3-day regimen of 200 mg fleroxacin once a day is effective in uncomplicated UTI, whereas in acute uncomplicated pyelonephritis, longer treatment is required. In complicated UTIs, if the drug is administered on an outpatient basis and Enterobacteriaceae predominate, 200 mg/day will usually effect a cure. In a hospital setting, or if gram-positive pathogens or *Pseudomonas* are the cause of the infection, 400 mg fleroxacin per day are necessary. Skin, soft tissue, bone and joint infections, as well as lower respiratory tract infections, including exacerbation of chronic bronchitis and nonpneumococcal pneumonia, are further indications that are amenable to successful treatment with fleroxacin.

Current clinical experience combined with excellent pharmacokinetic properties suggest that fleroxacin is a prominent drug within the group of newer fluorinated 4-quinolones.

References

1 Prosser BLT, Beskid G: Multicenter in vitro comparative study of fluoroquinolones against 25,129 gram-positive and gram-negative clinical isolates. Diagn Microbiol Infect Dis 1995;21:33–45.
2 Nightingale CH: Overview of the pharmacokinetics of fleroxacin. Am J Med 1993;94(suppl 3A):38S–43S.
3 Weidekamm E: Pharmacokinetics of fleroxacin in renal impairment. Am J Med 1993;94(suppl 3A):70S–74S.
4 Rohwedder C: Penetration of ciprofloxacin and metabolites into human lung, bronchial and pleural tissue after 250 and 500 mg oral ciprofloxacin. Chemotherapy 1991;37: 229–238.
5 Schmidt HGK, Neikes M, Weidenkamm E, Eggers H, Herbold D: Penetration of fleroxacin into bone and synovial fluid after administration of a single oral dose. Rev Infect Dis 1989;11(suppl 5):1268–1269.
6 Van Auwera P, Matsumoto T, Husson M: Intraphagocytic penetration of antibiotics. J Antimicrob Chemother 1988;22:185–192.
7 Sörgel F, Naber KG, Kinzig M, Müller C, Baumgartner U, Birner B, Savov O: Minor effect of ferrous sulphate on fleroxacin absorption (poster). 7th Eur Congr Clinical Microbiology and Infectious Diseases, Vienna, March 1995.
8 Cullmann W, Geddes AM, Weidekamm E, Urwyler H, Braunsteiner A: Fleroxacin: A review of its chemistry, microbiology, toxicology, pharmacokinetics, clinical efficacy and safety. Int J Antimicrob Agents 1993;2:203–230.
9 Iravani A: Multicenter study of single-dose and multiple-dose fleroxacin versus ciprofloxacin in the treatment of uncomplicated urinary tract infections. Am J Med 1993;94(suppl 3A):89–96.
10 Iravani A, St Clair P, Maladorno D: Fleroxacin in the treatment of uncomplicated urinary tract infections in women (abstract 725). 7th Eur Congr Clinical Microbiology and Infectious Diseases, Vienna, March 1995, p 140.
11 Pittman W, Moon JO, Hamrick LC, Cox CE, Clark J, Childs S, Pizzuti D, Fredericks J, St Clair P: Randomized double-blind trial of high- and low-dose fleroxacin versus norfloxacin for complicated urinary tract infection. Am J Med 1993;94 (suppl 3A):101S–104S.
12 Klimberg IW, Brooks RG, Finkelstein LH, Tice AD, Cox CE, Pittman WG: The safety and efficacy of fleroxacin once daily compared to twice daily ciprofloxacin in the treatment of complicated urinary tract infections (UTI). (poster 3150.) 19th International Congress of Chemotherapy, Montréal, July 1995.
13 Farkas SA: Intravenous fleroxacin versus ceftazidime in the treatment of acute nonpneumococcal lower respiratory tract infections. Am J Med 1993;94(suppl 3A):42S–49S.
14 Chodosh S: Efficacy of fleroxacin versus amoxicillin in exacerbations of chronic bronchitis. Am J Med 1993;94(suppl 3A):131S–135S.
15 Ulmer W: Fleroxacin versus amoxicillin in the treatment of acute exacerbation of chronic bronchitis. Am J Med 1993;94(suppl 3A):136S–141S.
16 Powers RD: Open trial of oral fleroxacin versus amoxicillin/clavulanate in the treatment of infections of skin and soft tissue. Am J Med 1993; 94(suppl 3A):155–158.
17 Tassler H: Comparative efficacy and safety of oral fleroxacin and amoxicillin/clavulanate potassium in skin and soft tissue infections. Am J Med 1993;94(suppl 3A):159–165.
18 Parish LC, Jungkind DL: Systemic antimicrobial therapy for skin and skin structure infections: Comparison of fleroxacin and ceftazidime. Am J Med 1993;94(suppl 3A): 166S–173S.
19 Butler T, Lolekha S, Rasidi C, Kadio A, del Rosa PL, Iskander H, Rubinstein E, Pastore G: Treatment of acute bacterial diarrhea: A multicenter international trial comparing placebo with fleroxacin given as a single dose or once daily for 3 days. Am J Med 1993;94(suppl 3A): 187S–194S.
20 Steffen R, Jori J, DuPont HL, Mathewson JJ, Stürchler D: Treatment of travellers' diarrhea with fleroxacin: A case study. J Antimicrob Chemother 1993;31:767–776.
21 Arnold K, Hong CS, Nelwan R, Zavala-Trujillo I, Kadio A, de Oliverira Barros MA, de Garis S: Randomized comparative study of fleroxacin and chloramphenicol in typhoid fever. Am J Med 1993;94:3A: 195S–200S.
22 Battilana C, Grados P, Delgado E, Carillo C: Comparative study of fleroxacin and tetracycline in the treatment of cholera; in Einhorn J et al (eds): Proceedings of the 18th International Congress of Chemotherapy, Stockholm, June 27–July 2, 1993. Recent Adv Chemother. Washington, American Society of Microbiology, 1994, pp 60–61.
23 Smith BL, Mogabgab WJ, Dalu ZA, Jones RB, Couglas JM, Hansfield HH, Hook EW III Jr, Viner BL, Schands JW Jr, McCormack WM: Multicenter trial of fleroxacin versus ceftriaxone in the treatment of uncomplicated gonorrhoea. Am J Med 1993;94(suppl 3A):81S–84S.
24 Green SL: Efficacy of oral fleroxacin in bone and joint infections. Am J Med 1993;94(suppl 3A):174.
25 Putz PA: A pilot study of oral fleroxacin given once daily in patients with bone and joint infections. Am J Med 1993;94:(suppl 3A):177.
26 Clinical Safety Data Pool. Basel, F. Hoffmann-La Roche Ltd., October 31, 1994.
27 Performance standards for antimicrobial susceptibility testing, NCCLS Document M100-S5, vol 14, No 16. Fifth Information Supplement, December 1994.

Chemotherapy

Chemotherapy 1996;42(suppl 1):10–16

Ross R. Bailey

Department of Nephrology, Christchurch Hospital, Christchurch, New Zealand

Single Oral Dose Treatment of Uncomplicated Urinary Tract Infection in Women

Case Presentation

Key Words
Urinary tract infection
Cystitis
Bacterial cystitis
Antibiotics
Quinolones
Fleroxacin

Abstract
A single oral dose of an antimicrobial agent is as effective as a conventional 3-day course for the treatment of uncomplicated lower urinary tract infections. A case history of a young woman with recurrent cystitis is presented to demonstrate this approach to the management of a common clinical problem. The place of antimicrobial prophylaxis is also discussed.

Introduction

There is now considerable evidence [1–5] that a single oral dose of an antimicrobial agent is as effective as a conventional short course for the treatment of uncomplicated lower urinary tract infections (UTIs).

The first drug evaluated in detail for single-dose therapy was amoxicillin given as a single 3-g oral dose [6]. A single dose of trimethoprim 600 mg or co-trimoxazole 1.92 g has been shown to be more effective than amoxicillin, and equivalent to a 3- to 5-day course of either preparation [3, 4]. Fosfomycin trometamol (3 g orally) has proven highly effective and is now marketed specifically for single-dose treatment of bacterial cystitis [5].

Recently, single doses of one of the 4-quinolones (e.g. norfloxacin, ciprofloxacin, fleroxacin, pefloxacin, lomefloxacin) have proven very effective [7], despite some doubt over their activity against *Staphylococcus saprophyticus.*

A case history is presented to demonstrate this approach to the management of a common clinical problem.

Case Presentation

A 20-year-old female university student presented with an 8-hour history of frequent micturition, together with dysuria, urgency, strangury, suprapubic discomfort and blood staining of the urine at the end of micturition. Over the past 6 months she had suffered

KARGER
E-Mail karger@karger.ch
Fax + 41 61 306 12 34

0009–3157/96/0427–0010$10.00/0

Ross R. Bailey, MD FRACP FRCP
Department of Nephrology
Christchurch Hospital
Private Bag 4710
Christchurch (New Zealand)

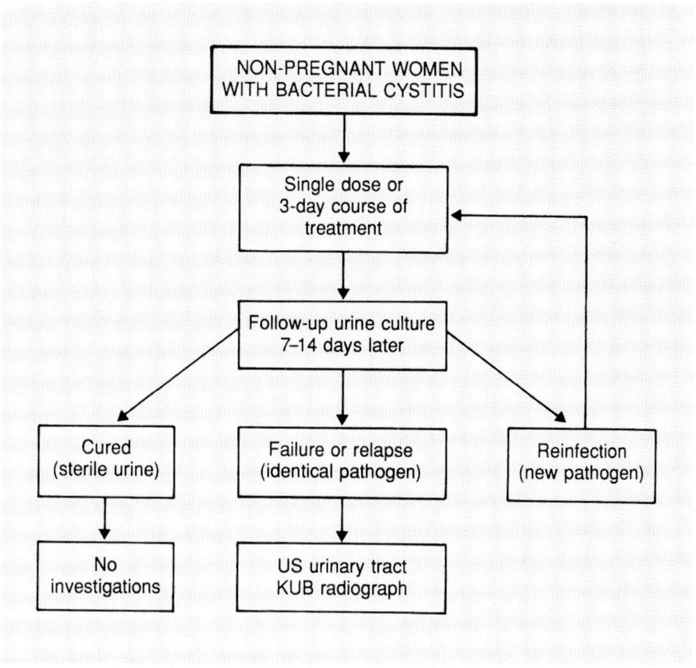

Fig. 1. Suggested scheme for the management of cystitis in women. KUB = kidney, ureter, bladder; US = ultrasonography [from *Clin Drug Invest* 1995; 9(suppl 1):8–13, with permission].

three similar episodes. She had had no urinary tract problems in childhood. For the past 6 months she had been taking an oral contraceptive preparation.

On examination, the patient appeared well with no fever or loin tenderness. The blood pressure was 136/90 mm Hg.

A midstream urine specimen showed 1+ protein and 2+ blood on dipstick analysis, while on microscopy there were >1,000 × 10^6 white cells/l, 50 × 10^6 red cells/l, no cellular casts and the urine grew *Escherichia coli* at a count of 80,000 c.f.u./ml. She was treated with a single 600-mg oral dose of trimethoprim.

Two months later the patient presented with another episode of cystitis and on this occasion the urine was infected with *S. saprophyticus.* She was treated with a single 800-mg oral dose of norfloxacin. The symptoms improved for a few days and then recurred. Seven days later the same organism was cultured. The patient was then treated with fleroxacin 200 mg daily for 3 days. She returned to the clinic 7 days after the last dose of fleroxacin and as the urine was sterile, low-dose antimicrobial prophylaxis was commenced.

Questions Arising from the Case Presented

What Was the Original Diagnosis?

This young woman presented with acute lower urinary tract symptoms and on examination of the urine, there was heavy pyuria and the specimen grew *E. coli.* The diagnosis was therefore one of bacterial cystitis.

How Would You Treat Her?

This patient could be treated with a single dose of an appropriate antimicrobial agent (fig. 1). Single-dose therapy can be considered as initial treatment for bacterial cystitis in sexually active women. There are many advantages of single-dose therapy, which have been fully documented elsewhere [1, 8]. The use of single-dose treatment has not predisposed to earlier, more severe, or more frequent recurrences.

Table 1. Single-dose oral regimens for uncomplicated UTIs

Trimethoprim 600 mg
Co-trimoxazole 1.92 g
Norfloxacin 800 mg
Ciprofloxacin 500 mg
Fleroxacin 400 mg
Lomefloxacin 400 mg
Pefloxacin 800 mg
Fosfomycin trometamol 3 g

Table 2. Drug regimens for an oral 3-day course of treatment for bacterial cystitis

Drug	Dose
Trimethoprim	300 mg o.d.
Co-trimoxazole	960 mg q 12 h
Nitrofurantoin	50 mg q 8 h
Norfloxacin	400 mg q 12 h
Ciprofloxacin	250 mg q 12 h
Fleroxacin	200 mg o.d.
Lomefloxacin	400 mg o.d.
Pefloxacin	400 mg q 12 h
Cephalexin	250 mg q 8 h
Cephradine	250 mg q 8 h
Cefaclor	250 mg q 8 h
Pivmecillinam	200 mg q 8 h
Amoxicillin	250 mg q 8 h
Augmentin	500 mg amoxicillin/ 125 mg clavulanic acid q 12 h

o.d. = Once daily; q = every.

An alternative to single-dose therapy in this clinical context is a short course of treatment (fig. 1). Many studies have demonstrated that there is no benefit in extending treatment beyond 3 days [9, 10]. Although some reviewers have indicated that a 3-day course of treatment may be more effective than single-dose therapy, this has not been substantiated in the literature and has been a conclusion reached by preferential citing of selected papers.

What Antimicrobial Agent(s) Would You Use?

Possible drugs for the single-dose oral treatment of cystitis in this young woman are included in table 1.

Alternatively, she could be treated with an oral 3-day drug regimen. There are many agents to choose from, some of which are included in table 2. There is no rationale for using two drugs in combination.

Are Any Further Investigations Necessary?

As this woman has had 3 similar episodes over the last 6 months, she is a candidate for a period of low-dose antimicrobial prophylaxis. Before commencing the latter, a strong case can be made for ultrasound examination of the urinary tract together with a plain abdominal radiograph. This would exclude urinary tract obstruction, calculi or gross renal scarring. There is no indication for cystoscopy in this clinical situation. Cystoscopy in younger women has been overused by urologists and in this large patient group rarely influences management.

We have shown that a woman who is treated successfully with single-dose therapy is unlikely to have a urinary tract abnormality [1, 6]. The practical corollary of this is that only those women who are single-dose failures warrant urinary tract investigation (fig. 1).

Is Post-Treatment Follow-Up Necessary?

The only way of clearly defining cure is to document a sterile urine specimen 7–14 days following treatment. The traditional recommendation has been a urine check 7 days after treatment, but some of the newer antimicrobial agents such as the long-acting quinolones (e.g. fleroxacin, lomefloxacin) are excreted in the urine for some time and it is therefore preferable to check for cure 10–14 days after completing treatment. A genuine bacteriological relapse will rarely occur if the urine is sterile at this stage. Any further recurrences will therefore be reinfections with a different pathogen. As a cost-saving approach, however, it has been suggested that for a patient such as the one under discussion, little is to be gained by culturing her urine if she is symptom-free [11]. Perhaps she should be instructed to return only if the symptoms failed to resolve, or recurred shortly after finishing treatment?

Why Did the Patient Fail Treatment with Norfloxacin?

When the patient returned with an episode due to *S. saprophyticus,* she was treated with a single oral dose of norfloxacin. In some studies [12, 13], single-dose therapy with a short-acting quinolone was less effective against *S. saprophyticus,* the second commonest pathogen causing cystitis in sexually active women.

What Prophylactic Treatment Regimen Would You Use?

Many women, such as the one discussed here, have recurrent symptomatic UTIs which cause considerable anxiety and morbidity. This is a particular problem in young, sexually active females. Many of these women are helped by completely emptying their bladders, particularly after intercourse, while some may benefit from applying an antiseptic cream (e.g. cetrimide 0.5%) to the periurethral area before intercourse.

If these simple techniques fail, the pattern of recurrences can be interrupted by instituting prophylactic therapy after the urine has been sterilized (fig. 2). Drug regimens for long-term, low-dose prophylaxis are included in table 3. Trials have shown that medication taken on alternate nights or even 3 nights a week is just as effective as nightly doses [14]. In some women, a dose after intercourse is equally effective [15, 16]. The initial period of prophylaxis should be for 6 months.

The excellent results with nitrofurantoin reflect the fact that the drug does not alter the fecal flora. The development of resistant bacterial strains during the administration of nitrofurantoin does not occur. Trimethoprim, co-trimoxazole and norfloxacin may have an advantage as prophylactic agents because they are excreted partly through the vagina, thus eliminating a possible source of infection.

Another approach for women who are having recurrent symptomatic UTIs is for selected women to use single-dose self-medication (fig. 2). In selected patients, single-dose self-therapy was shown to be efficacious and economical when compared with conventional treatment or conventional prophylaxis [17]. It is an increasingly common practice to encourage selected women with frequent episodes of bacterial cystitis who are unwilling to take long-term, low-dose prophylaxis to take a single dose or a 3-day course of a drug such as trimethoprim or fleroxacin when symptomatic (tables 1, 2), and to provide a urine specimen for culture 10–14 days later, or earlier, if the symptoms do not resolve within 3 days.

Comparative Clinical Studies of Single-Dose Quinolones for the Treatment of Bacterial Cystitis

There has been increasing interest in the use of the 4-quinolones in single-dose regimens for the treatment of bacterial cystitis. Bischoff [18] reported an 85% cure rate using norfloxacin in a single dose of 800 mg for the treatment of 50 women with bacterial cystitis. Rugendorff and Schneider [19] showed comparable cure rates in 59 women with cystitis allocated to treatment with norfloxacin in either a single 800-mg dose or 400 mg every 12 h for 3 days. Bischoff [18] later demonstrated that a single 400-mg dose of enoxacin was more effective (93%) and associated with

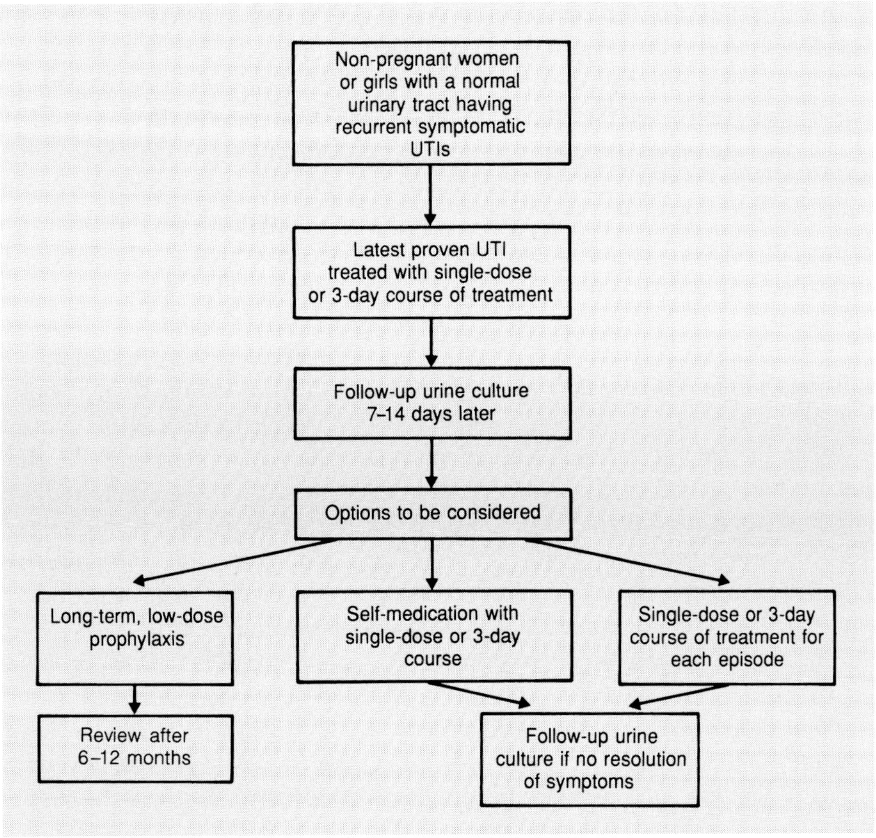

Fig. 2. Suggested scheme for the management of non-pregnant women with a normal urinary tract having recurrent symptomatic UTIs [from *Clin Drug Invest* 1995;9(suppl 1):8–13, with permission].

Table 3. Drug regimens for long-term, low-dose prophylaxis

Drug	Dose each night, alternate nights, 3 nights a weeks or after intercourse
Nitrofurantoin	50 mg
Trimethoprim	100 mg
Co-trimoxazole	240 mg
Norfloxacin	200 mg
Ciprofloxacin	125 mg
Cephalexin	125 mg (useful if renal insufficiency present)

fewer side effects than a single 3-g dose of amoxicillin (65%) for women with uncomplicated bacterial cystitis attending a urology clinic. Bailey et al. [12] subsequently showed in a study of women with cystitis in general practice that a single 400-mg dose of enoxacin was comparable to a single 600-mg dose of trimethoprim although enoxacin only eradicated 4 of 9 infections due to *S. saprophyticus*.

Since then, a number of comparative trials have been undertaken. Bailey [7] reviewed 17 such studies. In 9 of these, a single dose of a quinolone was compared with a single dose of another antimicrobial agent, and in 2 of these 9 studies, the quinolone was significantly better than the comparative drug. On both occasions, the latter was amoxicillin. In the other 7 studies, the quinolone and comparative drug were similar in efficacy. One of these 7 was a randomized trial comparing a single 400-mg dose of fleroxacin with a single 600-mg dose of trimethoprim in women presenting with acute lower urinary tract symptoms. Sixteen of 17 women with bacterial cystitis were cured with fleroxacin and 17 of 18 with trimethoprim. Side effects were minimal [20]. In the other 8 studies in this review [7], a single dose of a quinolone was compared with either a 3- or 7-day course of a comparative agent and in none of these studies was there a significant difference in efficacy. Bailey [7] concluded that single-dose quinolones were highly effective for the treatment of bacterial cystitis. The only concern would be their reported relative ineffectiveness against *S. saprophyticus* [12, 13, 21–23].

The fluoroquinolones have been a major advance in the management of uncomplicated UTIs and have a high in vitro activity against all major urinary tract pathogens.

Conclusion

Most UTIs in clinical practice present as bacterial cystitis in otherwise healthy women in the sexually active age group. The commonest pathogen is *E. coli* and the majority of the remainder are due to *S. saprophyticus*. Those women with cystitis which is uncomplicated (i.e. those who have a normal urinary tract and normal renal function) have a high cure rate when treated with any antimicrobial agent to which the infecting pathogen is sensitive [10, 24].

There is now a wealth of evidence to suggest that traditional dosage regimens for such lower UTIs are extravagant [25, 26]. Although it is still common practice for these patients to be treated with an antimicrobial agent for a prolonged period there is no convincing evidence that this is more effective than a single dose or a 3-day course of treatment. Ideally, any patient should be treated with the shortest course of the simplest, safest and cheapest antimicrobial agent that will eradicate the offending organism.

References

1 Bailey RR (ed): Single-Dose Therapy of Urinary Tract Infection, ed 1. Sydney, ADIS Health Science, 1983.

2 Bailey RR: Single-dose therapy for uncomplicated urinary tract infections. NZ Med J 1985;98:327–329.

3 Bailey RR: Single-dose therapy for uncomplicated urinary tract infections: An overview; in Kass EH, Svanborg Edén C (eds): Host-Parasite Interactions in Urinary Tract Infections. Chicago, University of Chicago Press, 1989, pp 405–410.

4 Bailey RR: Brief overview of single-dose therapy for uncomplicated urinary tract infections. Chemotherapy 1990;36(suppl 1):27–30.

5 Neu HC, Williams JD (eds): New Trends in Urinary Tract Infections: The Single-Dose Therapy. Basel, Karger, 1988.

6 Bailey RR, Abbott GD: Treatment of urinary-tract infection with a single dose of amoxicillin. Nephron 1977;18:316–320.

7 Bailey RR: Quinolones in the treatment of uncomplicated urinary tract infections. Int J Antimicrob Agents 1993;2:19–28.

8 Bailey RR: Management of lower urinary tract infections. Drugs 1993;45(suppl 3):139–144.

9 Hooton TM, Stamm WE: Management of acute uncomplicated urinary tract infection in adults. Med Clin N Am 1991;75:339–357.

10 Rubin RH, Beam TR Jr, Stamm WE: An approach to evaluating antibacterial agents in the treatment of urinary tract infection. Clin Infect Dis 1992;14(suppl 2):S246–S251.

11 Bailey RR: Cost-benefit considerations in the management of uncomplicated urinary tract infections in sexually active women. NZ Med J 1987;85:793–798.
12 Bailey RR, Gorrie SI, Peddie BA, Davies PR: Double-blind, randomised trial comparing single-dose enoxacin and trimethoprim for treatment of bacterial cystitis. NZ Med J 1987;100:618–619.
13 Garlando F, Rietiker S, Täuber MG, Flepp M, Meier B, Lüthy R: Single-dose ciprofloxacin at 100 versus 250 mg for treatment of uncomplicated urinary tract infections in women. Antimicrob Agents Chemother 1987;31:354–356.
14 Harding GKM, Buckwold DJ, Marrie TJ, Thompson L, Light RB, Ronald AR: Prophylaxis of recurrent urinary tract infection of female patients: Efficacy of low-dose, thrice-weekly therapy with trimethoprim-sulfamethoxazole. JAMA 1979;242: 1975–1977.
15 Stapleton A, Latham RH, Johnson C, Stamm WE: Postcoital antimicrobial prophylaxis for recurrent urinary tract infection: A randomized, double-blind, placebo controlled trial. JAMA 1990;264:703–706.
16 Pfau A, Sacks TG: Effective postcoital quinolone prophylaxis of recurrent urinary tract infections in women. J Urol 1994;152:136–138.
17 Wong ES, McKevitt M, Running K, Counts GW, Turck M, Stamm WE: Management of recurrent urinary tract infections with patient-administered single-dose therapy. Ann Intern Med 1985;102:302–307.
18 Bischoff W: Therapeutic concept in cystitis with norfloxacin: Single-shot therapy in acute cystitis (abstract). 14th Int Congr Chemotherapy, Kyoto, 1985, p 39–108.
19 Rugendorff EW, Schneider HJ: Randomized comparison of single-dose norfloxacin vs short-term therapy in acute lower urinary tract infections (abstract). 14th Int Congr Chemotherapy, Kyoto, 1985, p 39–107.
20 Bailey RR, Peddie BA, Davies PR: Randomized trial comparing single-dose fleroxacin (Quinodis) and trimethoprim (Triprim) for the treatment of bacterial cystitis. Clin Trials J 1989;26:411–416.
21 The Inter-Nordic Urinary Tract Infection Study Group: Double-blind comparison of 3-day versus 7-day treatment with norfloxacin in symptomatic urinary tract infections. Scand J Infect Dis 1988;20:619–624.
22 Piippo T, Pitkäjärvi T, Salo SA: Three-day versus seven-day treatment with norfloxacin in acute cystitis. Curr Ther Res 1990;47:644–653.
23 Neringer R, Forsgren A, Hansson C, Ode B: The South Swedish Lolex Study Group: Lomefloxacin versus norfloxacin in the treatment of uncomplicated urinary tract infections: Three-day versus seven-day treatment. Scand J Infect Dis 1992; 24:773–780.
24 Ohkoshi M, Kawada Y: Clinical Evaluation of Drug Efficacy in UTI. Amsterdam, Excerpta Medica, 1990.
25 Ohkoshi M, Naber KG (eds): Proceedings of the 2nd international symposium on clinical evaluation of drug efficacy in urinary tract infection. Infection 1992;20(suppl 3): S137–S243.
26 Ohkoshi M, Naber KG, Kawada Y (eds): Proceedings of the 3rd international symposium on clinical evaluation of drug efficacy in urinary tract infection. Infection 1994; 22(suppl 1):S3–S66.

Chemotherapy

Chemotherapy 1996;42(suppl 1):17–27

Fleroxacin in Complicated Urinary Tract Infections

Helen Giamarellou

1st Department of Propaedeutic Medicine, Athens University School of Medicine, Laiko General Hospital, Athens, Greece

Key Words
Fleroxacin, oral
Superinfections
Enterococci
Comparative studies
Intravenous
Urinary tract infection

Abstract
The clinical evaluation of antimicrobial agents for the treatment of urinary tract infections (UTIs) requires clinically useful categorization of patients plus concurrent laboratory diagnosis. Appropriate diagnostics include (1) proper collection techniques, (2) quantitative urine culture, (3) urinalysis to determine pyuria, and (4) radiological, urological and gynecological evaluation. Categorization of patients suffering from UTIs should take into consideration (1) the site of infection, (2) the clinical presentation, (3) the frequency of infections, and (4) the coexistence of complicating factors. UTIs in men and older women are mostly complicated. The characterization of a UTI as complicated or uncomplicated is important since in the former condition, multiresistant nosocomial microorganisms are frequently encountered. Experience with fleroxacin 200 or 400 mg once daily (o.d.) in >1,000 patients with complicated UTIs in noncomparative and comparative trials with other quinolones and cephalosporins has been very promising, with overall clinical and bacteriological cure rates of 86–95 and 89–95%, respectively. As with other quinolones, relapses and superinfections were mostly attributed to *Enterococcus, Staphylococcus* and *Pseudomonas* spp. The reported results make oral o.d. fleroxacin appropriate as an alternative to parenteral antimicrobials or for intravenous-to-oral switch therapy in complicated UTIs.

KARGER

E-Mail karger@karger.ch
Fax +41 61 306 12 34

0009–3157/96/0427–0017$10.00/0

Helen Giamarellou, MD
1st Department of Propaedeutic Medicine
Athens University School of Medicine
Laiko General Hospital
GR–11527 Athens (Greece)

Table 1. Complicated UTIs

One or more of the following present:
urinary calculi
cystic renal disease
obstruction
anatomical abnormalities
neurologic bladder dysfunction
foreign body
Urinary infections in patients with
diabetes mellitus
other metabolic or immunologic diseases
pregnancy
transplanted kidney
Infections that involve the prostate
Infections in nonambulatory elderly patients

Definitions

Evaluation of antimicrobials for urinary tract infections (UTIs) requires correct laboratory diagnosis and clinically useful categorization of patients. In the immunocompetent host, the former is based on the presence of (1) uropathogens appropriate to the infection, and (2) leukocytes in the urine corresponding to the inflammatory response. Bacteriuria without pyuria may represent colonization, as occurs in the elderly or in the presence of foreign bodies or infection in the neutropenic host.

The current established diagnostic criteria for complicated UTIs are shown in table 1 [1]. On the basis of such reported criteria, the categories proposed for antimicrobial agent evaluation, as defined by Schaeffer [2], are the following:

(1) women who are afebrile and have sporadic uncomplicated infections,

(2) women who are afebrile and have recurrent reinfections that are uncomplicated,

(3) women with febrile uncomplicated UTIs,

(4) women or men with complicated UTIs,

(5) men with chronic bacterial prostatitis, and

(6) men or women with symptomatic infections associated with indwelling catheters.

Complicated and uncomplicated UTIs are defined in terms of the diagnostic criteria shown in table 2. At the clinical level, a complicated UTI is equal to any combination of symptoms and male sex and/or urological abnormalities since UTIs in men are considered as complicated infections, while traditionally they occur at the extremes of age, i.e. in infants because of urinary anatomic abnormalities and in older men because of prostatic obstruction.

However, symptomatic UTIs may occur in young heterosexual circumcised men with apparently normal urinary tract as a result of heterosexual acquisition [3].

Complicated UTIs are common in clinical practice. Increases in the number of elderly and nursing-home residents and the high frequency of urinary catheterization and instrumentation in these populations will likely contribute to a rising incidence of complicated UTIs. Although *Escherichia coli* still accounts for 50% of infected episodes (vs. about 85% in community-acquired noncomplicated UTIs), the remaining patients are infected with difficult-to-treat bacterial pathogens, such as antibiotic-resistant gram-negative rods of hospital origin (e.g. *Pseudomonas aeruginosa, Klebsiella, Enterobacter, Citrobacter, Serratia* and *Providencia*) and gram-positive cocci *(Enterococcus, Staphylococcus epidermidis).*

It should be emphasized that by definition, complicated UTIs should have a sufficient anatomical or functional abnormality to cause significant obstruction [4]. Struvite and calcium apatite urinary calculi are the so-called infectious staghorn stones, classically representing one of the major causes of com-

Table 2. Differential diagnostic criteria for UTIs [1]

Type of UTI	Diagnostic criteria
Acute uncomplicated UTI in women	bacteriuria $\geq 10^3$ c.f.u./ml pyuria
Acute uncomplicated pyelonephritis	bacteriuria $\geq 10^4$ c.f.u./ml pyuria
UTI in men	bacteriuria $\geq 10^4$ c.f.u./ml pyuria
Complicated UTI (any sex)	bacteriuria $\geq 10^5$ c.f.u./ml pyuria
Asymptomatic bacteriuria	bacteriuria $\geq 10^5$ c.f.u./ml[a]

Pyuria = >5 leukocytes per high-power field in a centrifuged urine sample or, preferably, >10 leukocytes/mm in an unspun urine sample.
[a] In two consecutive midstream urine cultures >24 h apart plus pyuria.

plicated UTIs. The initial step in struvite calculogenesis is the adherence of ureolytic bacteria (e.g. *Proteus mirabilis, Klebsiella, Staphylococcus saprophyticus*) to Foley catheters, mucosal scars within the bladder or kidney and to the usual calcium stones. The alkaline pH produced (because the urease-forming bacteria split urea to ammonia) causes further precipitation of Mg^{2+} and Ca^{2+} perpetuating the size augmentation of the stone and the trapping of ureolytic bacteria as well.

Recently, the concept of bacterial biofilm production has further explained the pathogenesis of diagnosis and therapeutic difficulties with UTIs associated with catheters, indwelling ureteric stents, penile and external sphincter prostheses and struvite and calcium apatite stones. According to Nickel et al. [5], in the latter situation, there is an electrostatic interaction as well as chemical and hydrogen bonding between magnesium, ammonium phosphate and the biofilm matrix which binds these compounds and subsequently traps the struvite and calcium apatite crystals formed. Finally, crystals formed within the biofilm microenvironment adhere to the mucosa and coalesce into the initial stone nidus, which continues to grow with the addition of more bacteria, along with matrix production and subsequent crystal deposition. The complete removal of all stone fragments is absolutely necessary before or alongside antimicrobial therapy. Abnormalities that do not cause obstruction and stasis should not be regarded as complicating factors. While urinary tract instrumentation and pregnancy are well accepted as complicating conditions, diabetes and immunosuppressing conditions should be severe enough to compromise the host response.

Principles of Therapy

The rationale for treatment of UTIs is based on one or more of the following [6]:

(1) to reduce the duration and severity of symptoms,

(2) to prevent complications during pregnancy,

(3) to prevent complications during manipulation of the urinary tract,

(4) to prevent stone formation or growth,

(5) to prevent direct invasion of structures adjacent to the urinary tract such as the epididymis or the perinephric space,

(6) to prevent renal impairment, and

(7) to prevent infections that would make concomitant illnesses more serious, e.g. diabetes mellitus.

Because complicated UTIs vary in nature and severity, patients in this group represent a heterogenous category for which the diagnostic and therapeutic approaches taken should consider the patient as an individual. However, a general nonindividualized rule that represents the cornerstone in complicated UTI therapy is the surgical removal of the underlying obstruction of the urinary tract and of any foreign body, e.g. stones, catheters or prosthesis [4].

Among the several categories of patients with complicated UTIs, antimicrobial treatment is not recommended for asymptomatic bacteriuria in the elderly (>65 years) or in catheterized men and women, since it does not alter the natural history of the infection and does not prevent their complications. Additionally, prolonged courses of therapy frequently select multiply resistant pathogens that are difficult to eradicate [6].

Antimicrobial therapy will usually fail in patients with underlying urinary tract calculi. As a result, the therapeutic preference in these patients is to treat only symptomatic infections as if they were uncomplicated infections or in some instances to suppress persisting infections in patients whose calculi cannot be removed, particularly where there are urea-splitting organisms that promote stone growth [1, 6]. The permanent eradication of bacteria from the infected stone is also hindered because of the increased resistance to antibiotics of the 'sessile' biofilm bacteria, which can be attributed to (1) poor antibiotic penetration of the biofilm, and (2) the decreased metabolic activity of bacteria deep within the biofilm [5, 7].

Clinical Trials

Reviewing the evaluation of antibiotics in the treatment of UTIs, Norrby [8] has observed that 'the quality of studies performed and published has generally been less than optimal', despite the fact that UTI is a common infection and the main endpoint (elimination of bacteriuria) is easily assessed. Common errors in design and conduct of clinical trials, particularly in complicated UTIs, are [8]: heterogeneous patient samples presenting with mixed types of infection, various levels of infection, i.e. upper or lower urinary tract; underlying complicating factors and simultaneously performed surgical procedures; poorly defined risk groups; inadequate trial size to allow adequate testing of the null hypothesis; failure to account for all patients entered into a trial; incomplete laboratory procedures before, during and after therapy; type of superinfecting and relapsing bacteria not clarified, including their susceptibility patterns; lack of follow-ups, and inadequate methods for analysis of the results.

For complicated UTIs in particular, antimicrobial therapeutic problems arise from the underlying disease or anatomical problems that induce the infection and which are highly variable from patient to patient. A regimen effective against a complicated UTI may suppress bacterial growth during therapy rather than cure it, while rates of relapse and reinfection are, as a rule, high once therapy is discontinued. Furthermore, antimicrobial therapy can produce multiresistant pathogens which render therapeutic decisions even more difficult.

For an antibiotic to be considered successful in the treatment of complicated UTIs, the expected cure rates have been defined as follows [1]: >90% of patients with microorganism suppression and clinical improvement while receiving therapy (evaluation 3–5 days post-therapy) and >65% of patients with microorganism suppression and clinical improvement 5–9 days after completion of therapy. Four to six weeks after completion of therapy, follow-up information should be available for >50% of the patients evaluated 5–9 days after completion of therapy, with clinical cure and microorganism eradication documented in >40% of these patients. In general, the duration of therapy is recommended to last from 2–6 weeks.

The Choice of Antimicrobials for Complicated UTIs

Trimethoprim-sulfamethoxazole has been considered a successful antimicrobial when sensitive strains are present. Moreover, it is a suitable agent if long-term suppressive therapy is considered necessary. A second-generation cephalosporin or the administration of a β-lactamase inhibitor (since about one third of *E. coli* and all *Klebsiella* strains are nowadays β-lactamase producers) could be an appropriate choice [9]. However, the possible selection of more resistant clones should be considered because patients are usually pretreated several times with different classes of antimicrobials. In the latter case, an aminoglycoside (particularly when a parenteral compound is required) due to nausea and vomiting, an oral or parenteral third-generation cephalosporin (an extremely costly option), or an oral or parenteral fluoroquinolone should be considered.

The newer quinolones, in particular, possess several attractive properties for the therapy of complicated UTIs: (1) oral and parenteral formulations permit an easy intravenous-to-oral switch; (2) high urinary output as well as advantageous renal parenchymal concentrations; (3) advantageous kinetics in prostatic tissue; (4) safety and excellent tolerability permitting extra-prolonged therapy if required, and (5) their activity is not (significantly) influenced by the presence of a biofilm infection [7, 10, 11].

With biofilm infections, in particular, the following experimental evidence suggests that fluoroquinolones should be considered the drugs of choice. (1) In a device-related infection using a tissue/cage animal model, fleroxacin, ciprofloxacin, aztreonam and co-trimoxazole were tested against *E. coli* ATCC 25922. Co-trimoxazole, administered as prophylaxis or treatment, was inferior to the quinolones, despite susceptibility of the pathogens in vitro to all agents [10]. (2) In a rabbit model, fleroxacin demonstrated efficacy in the treatment of catheter-related infections: intravenous doses of 10–30 mg/kg every 8 h limited ascending infections of the urethra and bladder eliminating the catheter-associated biofilm and killed planktonic bacteria in the urine [11]. (3) In an in vitro agar diffusion model, the in vitro activity of ciprofloxacin, ofloxacin and pefloxacin in the presence of *Staphylococcus* slime was reduced by 9.5, 13 and 30%, respectively [7].

Experience with Fleroxacin in Complicated UTIs

Fleroxacin, a new trifluorinated 4-quinolone, displays excellent in vitro activity against the Enterobacteriaceae implicated in complicated UTIs. It has a bioavailability of almost 100% after oral administration, exhibits similar kinetic profiles in blood and urine following oral and parenteral administration,

Table 3. Noncomparative studies with fleroxacin

Reference	Dose mg	Duration days	Enrolled n	Bacteriological cure rate		Clinical cure rate	
				cured/ evaluated	%	cured/ evaluated	%
Cox (1988) [13]	400	7		21/22	95		
	600	7		22/22	100		
Bernstein-Hahn et al. (1989) [14]	200	7 or 10	28	24/28	86		
	400	7 or 10	29	24/29	83		
	600	7 or 10	28	25/28	89		
Cometta et al. (1989) [15]	400	7–14	27	23/25	92	23/25	92
Wolfhagen et al. (1990) [16]	200	7 or 10	20			12/14	86
	400	7 or 10	21			13/17	77
	600	7 or 10	21			10/14	71
Gelfand et al. (1993) [17]	400 i.v.–oral	7–14	68	26/32	81	27/32	84
Barclay et al. (1994) [18]	400 i.v.–oral	10	30	24/29	83		
DeGier (1994) [19]	400	7	26	14/18	78		
	i.v.–oral	14	28	12/16	75		

i.v. = Intravenous.

and its prolonged elimination half-life (9–13 h) permits once daily dosing. It is a safe quinolone with the most commonly reported adverse effects being insomnia (4%) and nausea (8%) [12].

Noncomparative Studies (table 3)

An early dose-finding study of fleroxacin in the treatment of complicated UTIs was reported by Bernstein-Hahn et al. [14]. They administered a single daily dose of 200, 400 or 600 mg fleroxacin at bedtime to 85 patients with undefined complicated UTIs for 10 days. Bacteriologic efficacy did not differ at the end of therapy, nevertheless, 5 weeks post-therapy, failures and relapses were more frequent in the group receiving 200 mg. However, the sample sizes in this study (28–29 patients per group) would be too small to demonstrate statistical significance. Taking the reported side effects into account, the authors recommended a single daily dose of 400 mg for 10 days. Wolfhagen et al. [16], in a similar double-blind, dose-finding study in 62 patients, reported a similar bacteriologic efficacy at 4–6 weeks post-therapy of 80%, but correlated the total number of adverse events with increasing dosage, and suggested, despite the low number of evaluable patients (45), that a 200- or 400-mg daily dose was preferable.

DeGier et al. [19], in a small study of 34 patients with poorly defined complicating underlying factors, evaluated the appropriate duration of therapy with fleroxacin: 7 versus 14 days. They concluded that combined intra-

Table 4. Comparative studies

Reference	Treatment and dose	Duration days	Enrolled n	Bacteriological cure rate		Clinical cure rate	
				cured/ evaluated	%	cured/ evaluated	%
Pittman et al. (1993) [23]	Flx 200 mg	10	102	68/71	96	61/71	86
	Flx 400 mg	10	97	56/61	92	58/61	95
	Nor 400 mg b.i.d.	10	97	52/58	90	50/58	86
Childs (1993) [24]	Flx 400 mg	10	152	92/94	98	90/94	96
	Nor 400 mg b.i.d.	10	149	85/96	89	80/96	84
Pummer (1993) [25]	Flx 400 mg	10	291	154/163	94		
	Nor 400 mg b.i.d.	10	296	150/163	92		
Naber and Sigl (1993) [26]	Flx 400 mg	7–14	39	24/30	80		
	Ofx 200 mg b.i.d.	7–14	41	32/35	91		
Palma et al. (1994) [27]	Flx 400 mg	7–14	27	25/27	93		
	Nor 400 mg b.i.d.	7–14	23	21/23	91		
Cox (1993) [20]	Flx 400 mg i.v.	4–21	320	155/165	94	142/165	86
	Cft 0.5–1 g t.i.d. or 1–2 g b.i.d.	4–21	154	78/82	95	73/82	89

Flx = Fleroxacin; Nor = norfloxacin; Ofx = ofloxacin; Cft = ceftazidime; b.i.d. = twice a day; i.v. = intravenous; t.i.d. = three times a day.

venous and oral therapy with fleroxacin is a satisfactory treatment modality, without any statistical difference in efficacy between 7 and 14 days. This result demonstrates the potential for fleroxacin in a short- (7-day) duration therapy, and is probably attributable to its long serum half-life.

Cox [20] administered intravenous fleroxacin 400 mg once daily for 4–21 days to 97 patients with pyelonephritis or signs and symptoms of UTI and complicating factors, with rates of clinical and bacteriologic cure at 5–9 days post-therapy discontinuation reaching 95%. Gelfand et al. [17] studied a sequential intravenous-oral switch regimen of 400 mg once daily for 7–14 days (intravenous for a mean of 3.2 days, oral for a mean of 5.3 days) in 32 evaluable patients with complicated UTIs or acute pyelonephritis among whom 4 were bacteremic. After a short-term follow-up, 81% were bacteriologically cured, staphylococci, enterococci and *Pseudomonas* strains representing the noneradicated species. This study indicated that fleroxacin is an effective and safe antimicrobial agent for sequential intravenous and oral therapy. However, *Pseudomonas* spp. and enterococci may be problematic pathogens, as has been reported with other fluoroquinolones [21, 22].

Comparative Studies with Other Quinolones (table 4)

The first study to include a large number of patients with complicated UTIs was that of Pittman et al. [23]. In a double-blind placebo-controlled multicenter study, they compared low- (200 mg) and high-dose (400 mg) fleroxacin once daily for 10 days with the standard norfloxacin therapy, 400 mg twice a day for 10 days in 296 patients suffering from a variety of complicated and/or recurring UTIs. All were symptomatic regarding frequency, dysuria, flank pain and systemic toxicity, while the presence of one or more of the following complicating factors was necessary for entry into the study: calculi, stricture, neurogenic bladder, post-prostatectomy status, prostatic obstruction, catheter- or instrument-induced infection and reflux. In the low-dose fleroxacin group, 68 (96%) of 71 patients at the 5- to 9-day post-therapy follow-up had bacteriologic cures compared with 56 (92%) of 61 in the high-dose fleroxacin group and 52 (90%) of 58 in the norfloxacin group. Interestingly, insomnia and nausea were reported in 33 patients given the high (400 mg) fleroxacin dose versus 10 patients treated with the low (200 mg) dose. The authors concluded that 'the efficacy and safety of low-dose fleroxacin are comparable to those of norfloxacin for treatment of complicated UTI'.

In a multicenter double-blind study, oral fleroxacin 400 mg once daily was compared to norfloxacin 400 mg twice daily for 10 days in the treatment of 301 patients suffering from serious UTIs [24]. Among 190 evaluable patients, 134 were considered to be complicated UTIs, based on well-defined underlying conditions. Bacteriologic results 5–9 days after completion of therapy (without distinguishing complicated from uncomplicated UTIs) were considered as cure in 98% of patients treated with fleroxacin, versus 89% receiving norfloxacin. The extremely high eradication rates, particularly with fleroxacin over a rather short duration of therapy, were attributed to the high fleroxacin levels obtained not only in urine but also in the affected tissue. However, this study as well as that of Pittman et al. [23] lacked prolonged follow-up (4–6 weeks post-therapy), which is necessary for the real evaluation of patients with underlying complicating diseases that render them vulnerable to relapses or reinfections [8].

Pummer [25] reported a large multicenter, double-blind, prospective, randomized comparative study in which oral fleroxacin 400 mg once daily was compared to norfloxacin 400 mg twice daily for 10 days in a total of 587 patients from 22 centers. Among them, 431 patients (217 fleroxacin, 214 norfloxacin) suffered from complicated UTIs, while, in addition, 13 versus 16%, respectively, also had diabetes mellitus. Without indicating patients' categories, bacteriologic results at short- (5–9 days) and long-term (4–6 weeks) follow-up were 94 versus 92% and 92 versus 84% in the two groups, respectively. However, the following should be pointed out: (1) only about 50% of the studied patients returned for the long-term follow-up; (2) the eradication rate for *P. aeruginosa* (75%) was clearly lower than for the other pathogens but was still comparable to that obtained with other new quinolones [28, 29]; (3) the occurrence of resistance was low (1%), while pathogens involved in superinfection, reinfection and relapse were susceptible to fleroxacin; (4) simultaneous urologic manipulations to resolve the complicating factors, and particularly to eliminate the obstruction of the urinary passages, were not clarified (i.e. if, and in how many patients, they were performed), and (5) adverse events, occurring in 26% of the fleroxacin group and in 14% of the norfloxacin group, were mainly related to the gastrointestinal tract and the central nervous system, and expressed mainly as nausea, insomnia and headache. It should be noted that the incidence of severe adverse events was low (1%), and rarely required discontinuation of therapy.

Naber and Sigl [26] in a very clear-cut although small study compared oral fleroxacin 400 mg once daily with oral ofloxacin 200 mg twice daily administered for 7–14 days in 80 patients with complicated UTIs who were, additionally, hospitalized urologic patients. Although bacteriologic results were evaluated only after a short-term follow-up, with cure rates of 80 and 91% for fleroxacin- and ofloxacin-treated patients, respectively, the authors pointed out that in both groups only gram-positive bacteria were involved in reinfections (5 vs. 1 patient, respectively) which were, in addition, resistant to quinolones. Therefore, although the number of observations was low, it is possible that significant emergence of resistant strains among enterococci, staphylococci and also *Pseudomonas* species can be expected if the quinolones in general are widely used in complicated UTIs which predispose to reinfections or relapses due to the underlying obstructing factors. In such situations, simultaneous surgical recovery of the urine flow should be a rule.

Fleroxacin, 400 mg once daily, was compared to twice daily norfloxacin 400 mg for 7 or 14 days in a small study of 50 patients with complicated UTIs, in whom the underlying factors rendering the UTIs complicated were not determined [27]. Bacteriological and clinical cure rates of over 90% were reported in both treatment groups.

Comparative Studies with β-Lactams

Intravenous fleroxacin, 400 mg once daily, was compared with intravenous ceftazidime, 0.5–2 g three times a day or 1–2 g twice a day, administered for 4–21 days for treatment of complicated UTIs [20] in 474 patients. The study was a multicenter, randomized, open-label trial with a 2:1 ratio of patients with pyelonephritis or signs and symptoms of UTI with complicating factors. Mainly as a result of protocol violations, pretherapy with antimicrobials <48 h before entering the study and sterile urine cultures, efficacy analysis included only 165 out of 320 fleroxacin-treated patients and 82 out of 154 ceftazidime-administered patients. Bacteriologic results at 5–9 days post-therapy discontinuation were very similar: 94 versus 95%, respectively. Interestingly, a total of 71 patients in the fleroxacin group and 27 in the ceftazidime group who had been cured at the short-term follow-up returned for long-term follow-up. In the fleroxacin group, 82% maintained cure versus 85% in the ceftazidime group. However, and despite the given definitions, the role of simultaneous urological manipulations was not clarified. In common with the Naber and Sigl study [26], most of the superinfections in both groups were attributed to *Enterococcus, Staphylococcus* and *Streptococcus* strains resistant to the preadministered antimicrobials. Despite the wide range of therapy duration (4–20 days), the rates of eradication of the most frequently isolated pathogens did not change as a function of treatment duration. The latter is a point of concern, since UTIs, if considered as complicated even after surgical removal of the urine flow obstruction, require a rational duration of antimicrobial therapy to kill the preexisting bacterial load in renal tissues. Possibly, the prolonged fleroxacin half-life renders a 5-day therapy with it equivalent to 8–10 days treatment with a β-lactam; however, whether short-term therapy with ceftazidime provides promising results is still an open question.

Conclusions

More than 1,800 patients with complicated UTIs have been included in several comparative and noncomparative studies with fleroxacin, among whom 1,050 were evaluable. *E. coli* was the isolated pathogen in 60–70% of the patients, *Klebsiella pneumoniae* in 10–20%, *P. aeruginosa* in 5–10%, *Pro-*

Table 5. Summary of published results with fleroxacin in complicated UTIs

Results of therapy	Number cured/total number treated			
	200 mg oral	400 mg oral	400 mg i.v.	400 mg i.v.–oral switch
Bacteriological cure rate	113/121 (93)	398/429 (93)	76/95 (80)	155/165 (94)
Clinical cure rate	73/85 (86)	334/360 (93)	27/32 (84)	142/165 (86)

Values in parentheses are percentages. i.v. = Intravenous.

teus spp. in 5–10% and several other species in 10–20%. A summary of the obtained and officially published results is given in table 5. Overall clinical cure rates with fleroxacin ranged from 84 to 96% and from 84 to 89% for the comparative antimicrobial agents, with bacteriological cure rates of 80–98% after 5–9 days versus 84–95%, reinfection rates of 3–11% versus 5–7% and relapse rates of 2–20% and 5–16%, respectively.

From these results, the following can be concluded. (1) Oral fleroxacin 200 or 400 mg once daily is as effective as other fluoroquinolones that require twice daily administration in the treatment of complicated UTIs. (2) Oral fleroxacin once daily is suitable for intravenous–oral switch therapy and is an appropriate alternative to other parenteral antibiotics. (3) As with other quinolones, the emergence of quinolone-resistant gram-positives and *P. aeruginosa* strains should be carefully monitored. The newer fluoroquinolones are useful in cases of nosocomial complicate, UTIs whenever multiresistant pathogens are implicated or suspected. Since fleroxacin is available in two dose sizes (200 and 400 mg), the appropriate regimen should be prescribed, taking into account risk-benefit considerations (i.e. superior tolerability of the lower dose for less severe infections versus superior efficacy of the higher dose for more severe infections).

References

1 Rubin RH, Shapiro ED, Andriole VT, Davis RJ, Stamm WE: Evaluation of new anti-infective drugs for the treatment of urinary tract infection. Clin Infect Dis 1992;15(suppl 1):216–227.
2 Schaeffer AJ: Diagnostics and categories of patients. Infection 1992;20 (suppl 3):138–142.
3 Krieger J, Ross SO, Simonsen J: Urinary tract infections in healthy university men. Antimicrob Agents Chemother 1993;149:1046–1048.
4 Bishop MC: Urosurgical management of urinary tract infection. J Antimicrob Chemother 1994;33 (suppl A):75–91.
5 Nickel JC, Costerton JW, McLean RJC, Olson M: Bacterial biofilms: Influence on the pathogenesis, diagnosis and treatment of urinary tract infections. J Antimicrob Chemother 1994;33(suppl A):31–42.
6 Ronald AR, Nicolle LE, Harding GKM: Standards of therapy for urinary infections in adults. Infection 1992;20(suppl 3):164–170.
7 Giamarellou H, Soulis M, Papapetropoulou M: Various antibiotic diffusion through slime producing *Staphylococcus* coagulase negative (CNS) clinical isolates (abstract 1310). 31st Interscience Conference on Antimicrobial Agents and Chemotherapy. Chicago, 1991.
8 Norrby JR: Evaluation of antibiotics for treatment of urinary tract infections. J Antimicrob Chemother 1994;33(suppl A):43–50.

9 Naber KG: Parenteral/oral switch in complicated urinary tract infection. Clin Drug Invest 1995;9(suppl 1): 14–19.
10 Widmer AF, Wiestner A, Frei R, Zimmerli W: Killing of non-growing and adherent *Escherichia coli* determines drug efficacy in device-related infections. Antimicrob Agents Chemother 1991;35:741–746.
11 Morck DW, Olson ME, McKay SG, Lam K, Prosser B, Cleeland R, Costerton JW: Therapeutic efficacy of fleroxacin for eliminating catheter-associated urinary tract infection on a rabbit model. Am J Med 1993;94 (suppl 3A):23–30.
12 Naber KG: Profile of fleroxacin. Clin Drug Invest 1995;9(suppl 1): 1–7.
13 Cox CE: An evaluation of two doses of Ro 23-6240 (fleroxacin) in the treatment of complicated or recurrent urinary tract infections (abstract 736). International Congress for Infectious Diseases. Rio de Janeiro, 1988.
14 Bernstein-Hahn L, Barclay CA, Iribarren MA, Casellas J, Traballi CA, Farinati A: Treatment of complicated urinary tract infection with fleroxacin: A dose-finding study. Rev Infect Dis 1989;11(suppl 5): 1359.
15 Cometta A, Kamm W, Berger JP, Glauser MP: Fleroxacin in noncomplicated urinary tract infections (UTI). 2nd International Symposium on New Quinolones. Geneva, 1988.
16 Wolfhagen MJAM, Hoepelman AIM, Verhoef J: Double-blind dose-range-finding study of fleroxacin for treatment of complicated urinary tract infections. Antimicrob Agents Chemother 1990;34:409–412.
17 Gelfand MS, Simmons BP, Craft RB, Grogan J, Amarshi N: A sequential study of intravenous and oral fleroxacin in the treatment of complicated urinary tract infection. Am J Med 1993;94(suppl 3A):126–130.
18 Barclay CA, Bernstein-Hahn L, Iribarren MA, Goldberg M: Intravenous fleroxacin followed by oral administration in the treatment of complicated UTI. 6th International Congress for Infectious Diseases. Prague, 1994.
19 DeGier RPE, Karperien A, Bouter PK, Zwinkels MGJ, Knol WRL, Verhoef J, Boon TA, Hoepelman AJM: Randomized study of combined intravenous oral fleroxacin for treatment of complicated urinary tract infections: 7 versus 14 days treatment duration. Proceedings of the 18th International Congress of Chemotherapy. Stockholm, 1994, pp 67–68.
20 Cox CE: Comparison of intravenous fleroxacin with ceftazidime for treatment of complicated urinary tract infection. Am J Med 1993;94 (suppl 3A):118–125.
21 Hooper DC, Wolfson JS: Fluoroquinolone antimicrobial agents. N Engl J Med 1991;324:384–394.
22 Nakanishi N, Yoshida S, Wakebe H, Inoue M, Mitsuhashi S: Mechanisms of clinical resistance to fluoroquinolones in *Enterococcus faecalis*. Antimicrob Agents Chemother 1991;35:1053–1059.
23 Pittman W, Moon JO, Hamrick LC, Cox CE, Clark J, Childs SJ, Pizzuti D, Fredericks J, Clair PS: Randomized double-blind trial of high- and low-dose fleroxacin versus norfloxacin for complicated urinary tract infection. Am J Med 1993;94(suppl 3A):101–104.
24 Childs SJ: Fleroxacin versus norfloxacin for oral treatment of serious urinary tract infections. Am J Med 1993;94(suppl 3A):105–107.
25 Pummer K: Fleroxacin versus norfloxacin in the treatment of urinary tract infections: A multicenter double-blind, prospective randomized comparative study. Am J Med 1993; 94(suppl 3A):108–113.
26 Naber KG, Sigl G: Fleroxacin versus ofloxacin in patients with complicated urinary tract infection: A controlled clinical study. Am J Med 1993;94(suppl 3A):114–117.
27 Palma P, Lima M, Netto N, Souza H: Comparative study of orally administered fleroxacin and norfloxacin in the treatment of complicated urinary tract infections. Proceedings of the 18th International Congress of Chemotherapy. Stockholm, 1994, p 74.
28 Leigh DA, Emmanuel FXS, Petch VJ: Ciprofloxacin therapy in complicated urinary tract infections caused by *Pseudomonas aeruginosa* and other resistant bacteria. J Antimicrob Chemother 1986;18(suppl D):117–121.
29 Williams AH, Gruneberg RN: Ciprofloxacin and cotrimoxazole in urinary tract infection. J Antimicrob Chemother 1986;18(suppl D):107–110.

Chemotherapy

Chemotherapy 1996;42(suppl 1):28–32

Thomas C. Gasser

Urologic Clinic, Kantonsspital Basel, Switzerland

Quinolone Prophylaxis in Transurethral Surgery

Key Words
Quinolones
Fleroxacin
Antimicrobial prophylaxis
Transurethral surgery

Abstract
In transurethral surgery, there is much controversy about the benefit of antimicrobial prophylaxis. Any urinary tract infection should be treated before proceeding with a transurethral operation. There is growing evidence that antimicrobial prophylaxis is also beneficial in cases with preoperatively sterile urine. A single dose or short-term prophylaxis appears to be sufficient. As newer fluoroquinolones reach very high tissue concentrations, have a suitable antimicrobial spectrum to cover most uropathogens and can be administered orally, these new drugs represent an attractive choice for prophylaxis in transurethral surgery.

Introduction

Since the pioneering animal studies of Burke [1], antimicrobial agents have been used for prophylactic purposes for many years and in all surgical fields. The need for antimicrobial prophylaxis is undisputed for many operations and clinical conditions such as cardiac, vascular and orthopedic surgery [2]. Conversely, in transurethral surgery, there is much controversy about the benefit of prophylactic administration of antimicrobial agents. Approximately 400,000 transurethral resections of the prostate (TURP) are performed each year in the US alone, making them the 10th most frequent of all operations, and they are clearly increasing in frequency [3]. Guidelines for the use and duration of antimicrobial prophylaxis are required to prevent unnecessary spending of precious healthcare resources.

The literature on this subject has therefore been reviewed, with special reference to the use of newer fluoroquinolones.

Problems of Antimicrobial Prophylaxis in Transurethral Surgery

In contrast to open surgery, the efficacy of antimicrobial prophylaxis in urologic surgery is difficult to assess, because there are no clear

KARGER
E-Mail karger@karger.ch
Fax + 41 61 306 12 34

0009–3157/96/0427–0028$10.00/0

Thomas C. Gasser, MD
Urologic Clinic, Department of Surgery
University Hospital (Kantonsspital)
Spitalstrasse 21
CH–4031 Basel (Switzerland)

definitions of postoperative infections such as wound infections or infection of a prosthesis. Moreover, the definition of significant bacteriuria has been debated and appears to be 10^3–10^4 c.f.u./ml, rather than the traditional level of 10^5 c.f.u./ml. Additionally, many of the complications are minor and, therefore, may be treated as they occur. Some urologists feel that it is safer to treat urinary tract infection when it occurs according to the infecting organism rather than to administer antimicrobials to every patient.

There must be a clear differentiation between patients with preoperatively infected urine and those with sterile urine. In cases of sterile urine, the term prophylaxis is appropriate. Conversely, when the urine is infected, administration of antimicrobials should be referred to as therapy.

Preoperatively Infected Urine

In a study of more than 1,000 patients who underwent TURP, infectious complications such as urinary tract infection and septicemia were the most common [4]. Possible sources of infection include the inevitable lens-eye contact, the urethra with its wide variety of bacteria, the prostate and possibly contaminated irrigation fluids [5]. However, with the implementation of video cameras and television screens to perform TURP, lens-eye contact is no longer necessary and infection rates may fall.

While bacteremia following TURP is a common finding, occurring in 11% of patients with sterile urine, this number rises up to 50% in cases of preoperatively infected urine [6, 7]. Preoperative bacteriuria is a clear risk factor for postoperative septicemia [8]. It is therefore generally agreed that urinary tract infections should be treated to clear the urine before TURP is performed [2].

However, even when every reasonable effort is taken to identify bacteriuric patients preoperatively, up to 19% will unknowingly be infected during surgery [9]. Patients with a preoperative indwelling catheter are at particularly high risk of being bacteriuric. In a study that included these unsuspected cases, the treated patients fared significantly better than the nontreated patients with respect to febrile episodes, complications and length of hospital stay [10].

Preoperatively Sterile Urine

The question whether antimicrobials should be administered to patients with sterile urine who undergo TURP is much discussed in the literature. In a 1979 review article, Chodak and Plaut [11] were unable to draw definite conclusions as to the benefit of antimicrobial prophylaxis in urologic surgery for this group of patients. They criticized the poor design of the studies published up to that date and called for prospective, randomized and controlled trials. Since then, many such studies have been published, but they have not completely ended the dispute. Table 1 lists selected studies of antimicrobial prophylaxis published within the last 10 years. While inconsistencies exist in the design of these studies, all patients had sterile urine preoperatively, were evaluated between the 5th and 7th postoperative day and all studies were controlled versus no antimicrobials or placebo.

Three studies found no effect of prophylaxis on the infection rate [14, 16, 18]. This may well be for statistical reasons. As we have calculated previously, due to the β error, these studies had a statistical power of <50%, indicating a 50% chance of overlooking a significant difference between treatment and placebo groups [24]. The study by Millar et al. [14] was well designed and found no significant

Table 1. Controlled studies (placebo or no antimicrobial) of antimicrobial prophylaxis in patients undergoing TURP

Year	Series	Level of significant bacteriuria c.f.u./ml	Number of patients	Prophylaxis	Rate of infection, %	
					treated patients	control patients
1985	Allan and Kumar [12]	N/A	100	mezlozillin[a]	20	64*
1986	Prokocimer et al. [13]	$>10^5$	90	cefotaxime[b]	4	30*
1987	Millar et al. [14]	$>10^5$	179	aztreonam[a]	22	32
1988	Desai et al. [15]	$>10^5$	80	enoxacin[b]	8	38*
1988	Stricker and Grant [16]	$>10^5$	93	ampicillin/gentamicin[a]	17	16
1988	Taylor et al. [17]	$>10^5$	235	temocillin[b]	13	24*
1989	Houle et al. [18]	$>10^5$	110	cefoperazone[b]	0	2
1989	Abbou et al. [19]	$>10^5$	80	pefloxacin[b]	5	33*
1991	Ueda et al. [20]	$>10^4$	58	ofloxacin[b]	27	56*
1992	Slavis et al. [21]	$>10^4$	107	cefonicid[a]	12	37*
1993	Viitanen et al. [22]	$>10^5$	396	ceftriaxone[a]	8	22*
1993	Hargreave et al. [23]	$>10^4$	517	ceftazidime[b]	12	34*
1993	Gasser et al. [unpubl. data]	$>10^4$	75	fleroxacin[b]	3	23*

N/A = Information not available; * = treated and control patients differ significantly.
[a] Administered preoperatively.
[b] Administered perioperatively (before, during, and after surgery).

difference between the two groups. Stricker and Grant [16] found that a break in the closed drainage system was far more important than the use of antimicrobials.The study by Houle et al. [18] found an unmatched low bacteriuria number, even in the control group, which was not explained.

All the other studies demonstrated statistically significant reductions in infection rate from approximately 30% to well below 10%. On the basis of these studies antimicrobial prophylaxis is strongly advised for patients undergoing transurethral surgery even when preoperatively the urine is sterile.

The optimal duration of prophylaxis cannot be determined from the literature. There is much variation as to the introduction and timing of prophylaxis as well as the route of drug administration. In one study, single-dose prophylaxis was clearly superior to placebo [21]. Other investigators found single-dose prophylaxis to be as effective as a 3-day regimen [25]. In one controlled study from Israel (ceftriaxone versus no antimicrobial), the drug was administered 1 h prior to surgery and a second dose 1 h prior to removal of the catheter. In the ceftriaxone group, the bacteriuria rate was significantly reduced, febrile episodes were fewer and the hospital stay shorter [26]. This may be an indication that a short antibiotic cover, possibly until removal of the catheter, may be beneficial.

At our institution, short antimicrobial prophylaxis is administered starting immediately prior to surgery until removal of the catheter (usually 24 h after surgery).

Role of Quinolones in Transurethral Surgery

The unique mode of action of newer fluoroquinolones is the binding and blocking of the bacterial DNA gyrase enzyme. The broad antimicrobial spectrum covers gram-positive and gram-negative bacteria including several *Pseudomonas* strains and β-lactam bacteria. Most of the uropathogens are thus covered [27]. All fluoroquinolones are administered orally and some intravenously as well. Quinolones reach very high serum, tissue and urine concentrations. For example, fleroxacin reaches prostatic tissue concentrations 1.5-fold higher than the simultaneous plasma concentration and the tissue level was severalfold higher than the MICs for most urinary tract pathogens. The half-life of most quinolones has been calculated to be between 3 and 5 h, except for fleroxacin and pefloxacin, which have half-lives of 11.2 and 10.5 h, respectively, allowing once-daily administration [28]. Quinolones are usually well tolerated, also in elderly people who represent the largest population in transurethral surgery. To date, the published studies comparing quinolones and cephalosporins as prophylactic agents have indicated equal drug efficacy. In one study of 207 patients, oral lomefloxacin was equally effective as parenteral cefotaxime in preventing postoperative infections [29]. In a smaller study of 63 patients, a single oral lomefloxacin dose was as efficacious and as safe as a single intravenous dose of cefuroxime in preventing postoperative urinary tract infection [30]. In a recent randomized, placebo-controlled study at our institution of 75 patients undergoing transurethral surgery, oral fleroxacin significantly reduced the postoperative bacteriuria rate from 23 to 3% [unpubl. data].

Discussion and Conclusion

There is still much controversy as to the benefit of antimicrobial prophylaxis in transurethral surgery. There is general agreement to treat any urinary tract infection until it is cleared before proceeding with a transurethral resection of the prostate.

There is growing evidence that prophylactic administration of antimicrobial agents can prevent infectious complications even in the face of preoperatively sterile urine, although not all authors would agree with this. Numerous well-designed, prospective, randomized and controlled (placebo or no antibiotic) studies have clearly shown a benefit for the prophylaxis group. The danger for patients with unsuspected and undetected urinary tract infection before transurethral surgery should not be underestimated. These patients particularly benefit from antimicrobial administration intended as prophylaxis, but in fact given as treatment.

The newer fluoroquinolones have been shown to be a very effective, well-tolerated and safe group of new antibiotics. They can be administered orally and some of them even only once a day (such as fleroxacin). Therefore, the antimicrobial activity, the pharmacokinetic properties, the high tissue concentration and their safety make the quinolones an attractive and cost-effective choice for prophylaxis in transurethral surgery.

Acknowledgment

The help of M. Luaces in preparing the manuscript is gratefully acknowledged.

References

1 Burke JF: The effective period of preventive antibiotic action in experimental incisions and dermal lesions. Surgery 1961;50:161–168.
2 Kaiser AB: Antimicrobial prophylaxis in surgery. N Engl J Med 1986; 315:1129–1138.
3 Rutkow IM: Urological operations in the United States: 1979 to 1984. J Urol 1986;135:1206–1208.
4 Paquin JM, Perreault JP, Faucher R, Mauffette F, Valiquette L, Lapointe S: Complications immédiates de la résection transurétrale de la prostate: Étude de 1000 cas consécutifs. Can J Surg 1988;31:438–440.
5 Larsen EH, Gasser TC, Madsen PO: Antimicrobial prophylaxis in urologic surgery. Urol Clin North Am 1986;13:591–604.
6 Creevy CD, Feeny MJ: Routine use of antibiotics in transurethral prostatic resection: A clinical investigation. J Urol 1954;71:615–623.
7 Sullivan NM, Sutter VL, Mims MM, Marsh VH, Finegold SM: Clinical aspects of bacteremia after manipulation of the genitourinary tract. J Infect Dis 1973;127:49–55.
8 Cafferkey MT, Falkiner FR, Gillespie WA, Murphy DM: Antibiotics for the prevention of septicaemia in urology. J Antimicrob Chemother 1982;9:471–477.
9 Shearman CP: Antibiotic prophylaxis in transurethral prostatectomy. Br J Surg 1987;74:653.
10 Hargreave TB, Hindmarsh JR, Elton R, Chisholm GD, Gould JC: Short-term prophylaxis with cefotaxime for prostatic surgery. Br Med J 1982;284:1008–1010.
11 Chodak GW, Plaut ME: Systemic antibiotics for prophylaxis in urologic surgery: A critical review. J Urol 1979;121:695–699.
12 Allan WR, Kumar A: Prophylactic mezlocillin for transurethral prostatectomy. Br J Urol 1985;57:46–49.
13 Prokocimer P, Quazza M, Gibert C, Lemoine JE, Joly ML, Dureuil B, Moulonguet A, Manuel C, Desmonts JM: Short-term prophylactic antibiotics in patients undergoing prostatectomy: Report of a double-blind randomized trial with 2 intravenous doses of cefotaxime. J Urol 1986;135:60–64.
14 Millar MR, Inglis T, Ewing R, Clark P, Williams RE, Lacey RW: Double-blind study comparing aztreonam with placebo for prophylaxis of infection following prostatic surgery. Br J Urol 1987;60:345–348.
15 Desai KM, Abrams PH, White LO: A double-blind comparative trial of short-term orally administered enoxacin in the prevention of urinary infection after elective transurethral prostatectomy: A clinical and pharmacokinetic study. J Urol 1988;139:1232–1235.
16 Stricker PD, Grant ABF: Relative value of antibiotics and catheter care in the prevention of urinary tract infection after transurethral prostatic resection. Br J Urol 1988; 61:494–497.
17 Taylor EW, Lindsay G, West of Scotland Surgical Infection Study Group: Antibiotic prophylaxis in transurethral resection of the prostate with reference to the influence of preoperative catheterization. J Hosp Infect 1988;12:75–83.
18 Houle AM, Mokhless I, Sarto N, Elhilali MM: Perioperative antibiotic prophylaxis for transurethral resection of the prostate: Is it justifiable? J Urol 1989;142:317–319.
19 Abbou C, Chopin D, Nguyen M, Theodon P, Antiphon P, Goulois R, Bouleau D: Short-term pefloxacin prophylaxis in transurethral resection of the prostate. Rev Infect Dis 1989;11(suppl 5):1361.
20 Ueda S, Matsuoka K, Yamashita T, Yoshitake N, Yoshizumi O, Noguchi M, Hayashi K, Noda S, Eto K: Antibiotic prophylaxis with ofloxacin in patients undergoing transurethral prostatectomy. Eur J Clin Microbiol Infect Dis 1991;(special issue):322–323.
21 Slavis AS, Miller JB, Golji H, Dunshee CJ: Comparison of single-dose antibiotic prophylaxis in uncomplicated transurethral resection of the prostate. J Urol 1992;147:1303–1306.
22 Viitanen J, Talja M, Jussila E, Nurmi M, Permi J, Puolakka VM, Rintala E, Salmela H, Tiitinen J, Tuhkanen K: Randomized controlled study of chemoprophylaxis in transurethral prostatectomy. J Urol 1993; 150:1715–1717.
23 Hargreave TB, Botto H, Rikken GH, Hindmarsh JR, McDermott TE, Mjolnerod OK, Petays P, Schalkhauser K, Stellos A: European collaborative study of antibiotic prophylaxis for transurethral resection of the prostate. Eur Urol 1993;23: 437–443.
24 Gasser TC, Graversen PH, Madsen PO: Antimicrobial prophylaxis in surgery. N Engl J Med 1987;316: 1089.
25 Bijl W, Janknegt RA: Single-dose versus 3-day prophylaxis with ciprofloxacin in transurethral surgery: A clinical trial. Urol Int 1993;51:73–78.
26 Raz R, Almog D, Elhanan G, Shental J: The use of ceftriaxone in the prevention of urinary tract infection in patients undergoing transurethral resection of the prostate (TUR-P). Infection 1994;22:347–349.
27 Hooper DC, Wolfson JS: Fluoroquinolone antimicrobial agents. N Engl J Med 1991;324:384–395.
28 Gasser TC, Graversen PH, Madsen PO: Fleroxacin (Ro 23-6240) distribution in canine prostatic tissue and fluids. Antimicrob Agents Chemother 1987;31:1010–1013.
29 Klimberg IW, Childs SJ, Madore RJ, Klimberg SR: A multicenter comparison of oral lomefloxacin versus parenteral cefotaxime as prophylactic agents in transurethral surgery. Am J Med 1992;92:121S–125S.
30 Charton M, Mombet A, Gattegno B: Urinary tract infection prophylaxis in transurethral surgery: Oral lomefloxacin versus parenteral cefuroxime. Am J Med 1992;92:118S–120S.

Chemotherapy

Chemotherapy 1996;42(suppl 1):33–42

Günter Marklein

Institute for Medical Microbiology and Immunology, Rheinische Friedrich-Wilhelms-Universität Bonn, Germany

Quinolones in Everyday Clinical Practice: Respiratory Tract Infections and Nosocomial Pneumonia

Key Words
Fleroxacin
Ciprofloxacin
Pneumonia
Nosocomial
Community-acquired
Parenteral-oral sequential therapy

Abstract
Currently available fluoroquinolones have established their value in the treatment of lower respiratory tract infections due to gram-negative rods and *Staphylococcus aureus*. The fact that these drugs are absorbed (and well tolerated) when given orally is a major positive feature. The once daily dosage of fleroxacin [400 mg once daily intravenously (i.v.) for 2–4 days followed by oral doses of 400 mg for up to 10 days] was compared with twice daily ciprofloxacin (400 mg twice daily i.v. for 2–4 days followed by oral doses of 2 × 500 mg for up to 10 days) for treatment of inpatients with pneumonia confirmed by clinical signs and chest X ray. To date, 93 evaluable patients have been enrolled in this study. Clinical cure and improvement rates were 73.3% in the fleroxacin group and 79.2% in the ciprofloxacin group. The rate of adverse clinical or laboratory events was similar in both study groups.

Introduction and Overview

In general, therapeutic regimens for bacterial diseases of the lower respiratory tract must be selected empirically, on the basis of the spectrum of infectious agents expected in specific situations. Treatment is usually initiated before microbiological results are available and a causative diagnosis is often complicated by the fact that bacteriological cultures of respiratory tract specimens may yield an array of facultatively pathogenic microorganisms. On the other hand, acute bacterial exacerbations of chronic bronchitis (AECB) and pulmonary infections are major indications for the administration of antimicrobial agents providing high rates of benefit.

Leading pathogens of community-acquired and nosocomial pneumonia are compiled in table 1. While in community-ac-

KARGER

E-Mail karger@karger.ch
Fax +41 61 306 12 34

0009–3157/96/0427–0033$10.00/0

Günter Marklein, MD
Institute for Medical Microbiology and Immunology of RFWU
Sigmund-Freud-Strasse 25
D–53127 Bonn (Germany)

Table 1. Predominant etiological agents of community-acquired and nosocomial pneumonia

Community-acquired pneumonia	Nosocomial pneumonia
Streptococcus pneumoniae	*Staphylococcus aureus*
Haemophilus influenzae	*Klebsiella pneumoniae*
Mycoplasma pneumoniae	*Escherichia coli*
Chlamydia pneumoniae	Various Enterobacteriaceae
Legionella pneumophila	*Pseudomonas aeruginosa*
Moraxella catarrhalis	*Legionella pneumophila*

quired pneumonia leading pathogens are *Streptococcus pneumoniae, Haemophilus influenzae,* and *Mycoplasma pneumoniae,* nosocomial pneumonia is mainly caused by gram-negative bacteria in well over 50% of cases, Enterobacteriaceae, for example *Escherichia coli* or *Klebsiella pneumoniae* and *Pseudomonas aeruginosa,* being the main pathogens.

Because there is no reliable method to distinguish promptly between pneumonia caused by the agents of 'typical' acute bacterial disease and 'atypical' pneumonia, empiric antibiotic therapy of patients with pneumonia is difficult to establish.

In hospitalized patients with severe community-acquired pneumonia, antibacterial therapy before culture results often comprises a second-generation cephalosporin (e.g. cefuroxime or cefamandole) or a β-lactam/β-lactamase inhibitor combination (e.g. ampicillin/sulbactam or amoxicillin/clavulanic acid) with erythromycin added for *Legionella* coverage.

Elderly patients, especially those from nursing homes, or patients with significant underlying disease may have pneumonia caused by aerobic gram-negative rods. In such cases, reasonable regimens of antibiotic therapy would be combinations with third-generation cephalosporins (cefotaxime or ceftriaxone, or ceftazidime for infection with *P. aeruginosa*) or an extended-spectrum penicillin/β-lactamase inhibitor combination (piperacillin/tazobactam), all plus an aminoglycoside or a quinolone.

Before treatment regimens for nosocomial infections of the lower respiratory tract can be selected, several issues need to be considered: (1) the preceding antibiotic therapy that might have selected for more resistant organisms, (2) underlying chronic bronchitis that would increase the risk of *H. influenzae* infection, (3) the recent epidemiological experience in a given hospital or intensive care unit (e.g. a high incidence of multiresistant *Acinetobacter, Enterobacter,* or *Serratia* strains, and (4) potential pathogens with unusual antimicrobial resistance patterns known from prior surveillance cultures of the patient's bronchial secretions.

For the past few years, quinolones have been widely prescribed as therapy for AECB as well as for both community- and nosocomially acquired respiratory tract infections, not least as a consequence of the growing importance of β-lactam resistance due to β-lactamase-producing bacteria [1, 2].

The antibacterial spectrum of the quinolones as a group covers mainly gram-negative, aerobically growing bacteria at very low MICs of less than 0.1 mg/l for most of the important pathogens (table 2); gram-positive bacteria, with the exception of *Staphylococcus aureus,* are inhibited only by higher levels of most quinolones [3]. Sparfloxacin is effective

Table 2. Antibacterial activity of some fluoroquinolones against aetiological agents of respiratory tract infections (data from the literature)

Bacterial species	Fleroxacin	Ciprofloxacin	Ofloxacin	Sparfloxacin
Gram-negative organisms				
Haemophilus influenzae	0.06	≤0.06	≤0.06	≤0.06
Klebsiella spp.	0.5	0.125	0.5	0.25
Escherichia coli	0.125	≤0.06	0.125	0.06
Enterobacter spp.	0.125	0.125	0.5	0.25
Serratia spp.	1.0	0.25	1.0	1.0
Pseudomonas aeruginosa	4.0	0.50	4.0	2.0
Gram-positive and other organisms				
Streptococcus pneumoniae	8.0	2.0	2.0	0.25
Staphylococcus aureus	0.5	1.0	0.5	0.25
Streptococcus pyogenes	8.0	2.0	2.0	0.5
Mycoplasma pneumoniae	4.0	3.0	2.0	0.25
Legionella spp.	0.25	0.06	0.06	0.06
Chlamydia pneumoniae	0.5	0.25	1.0	1.0

Values are the MIC_{90} in mg/l.

against *S. pneumoniae* and the majority of other agents of community-acquired pneumonia (table 3).

The following overview will focus on the question: are there clinical data to support the efficacy of ciprofloxacin and fleroxacin in the treatment of lower respiratory tract infections and pneumonia?

In clinical trials involving patients with pneumonia and other lower respiratory tract infections, the clinical efficacy rates of ciprofloxacin compared favorably with those of amoxicillin [4] and imipenem [2] and were superior to that of pefloxacin [6].

In a prospective, randomized, double-blind multicenter trial by Fink et al. [7], intravenously administered ciprofloxacin (400 mg every 8 h) was compared with imipenem (1 g every 8 h) for the treatment of severe pneumonia. Among 405 patients, 79% required mechanical ventilation and 78% had nosocomial pneumonia. Ciprofloxacin-treated patients had a higher bacteriological eradication rate (69%) than imipenem-treated patients (59%). They also had a higher clinical response rate (69 vs. 56%). The greatest difference between the treatment groups was in eradication of members of the Enterobacteriaceae family (ciprofloxacin 93%, imipenem 65%). When *P. aeruginosa* was recovered from initial respiratory tract cultures, failure to achieve bacteriological eradication was common in both treatment groups (67% for ciprofloxacin and 59% for imipenem). Development of resistant *P. aeruginosa* during therapy occurred commonly in both groups (ciprofloxacin 33%, imipenem 53%). The authors concluded that monotherapy with ciprofloxacin or imipenem for severe pneumonia is a safe and effective initial strategy which needs to be modified if *P. aeruginosa* is suspected or recovered as a causative agent.

In two prospective, randomized clinical trials with fleroxacin, by Chodosh [4] and

Table 3. Fleroxacin clinical studies – published results with AECB and acute nonpneumococcal lower respiratory tract infection (ANPLRTI)

Indication	Design	Treatment	Duration days	Enrolled	Cure rates				Adverse events	
					clinical		bacteriological			
					cured/ evaluated	%	cured/ evaluated	%	patients/ evaluated	%
AECB [4, 5]	double-blind placebo group multicenter	fleroxacin 400 mg o.d.	7	417	131/144	91	114/146	97	103/145	25
		amoxicillin 500 mg t.i.d.	7	404	124/154	81	122/156	78	41/402	10
ANPLRTI [6]	open label placebo group multicenter	fleroxacin 400 mg i.v. o.d.	4–21	209	59/67	88	58/68	85	24/206	12
		ceftazidime 1–2 g b.i.d., 0.5–2 g t.i.d.	4–21	110	40/49	82	40/49	82	15/106	13

o.d. = Once daily; t.i.d. = three times a day; i.v. = intravenous; b.i.d. = twice daily.

Ulmer [5], conducted according to the same study protocol, more than 400 patients suffering from AECB were enrolled (table 3). The efficacy and safety of 400 mg fleroxacin once daily given orally for 7 days were compared to amoxicillin 500 mg three times daily. In the study published by Farkas [6], fleroxacin 400 mg given intravenously once daily was compared with ceftazidime plus erythromycin in patients with acute nonpneumococcal lower respiratory tract infection (table 3). The clinical cure rates of fleroxacin were 91 and 88% (table 3), in comparison to 81% for amoxicillin and 82% for ceftazidime. Bacteriological cure rates for fleroxacin were 97 and 85% (table 3). Compared to amoxicillin, the rate of adverse events was markedly higher in the fleroxacin group (25 vs. 10%), but in comparison to ceftazidime, the rate of adverse events was in the same range (12 vs. 13%; table 3).

Patients and Methods

The study to be described here was carried out as a prospective, randomized, double-blind multicenter trial. Adult patients of either sex were recruited from inpatient populations of twenty medical centers distributed through Germany.

Patients <18 years of age and pregnant or nursing women were excluded. Further exclusion criteria were hypersensitivity to quinolones and evidence of impaired hepatic and renal function. Patients who fulfilled the clinical inclusion criteria (table 4) were assigned, according to randomization lists, to fleroxacin or ciprofloxacin in a 1:1 ratio. The intravenous dosage was 400 mg fleroxacin once daily or 400 mg ciprofloxacin twice daily for up to 4 days, followed by oral administration of fleroxacin 400 mg once daily or ciprofloxacin 500 mg twice daily.

The total duration of treatment was up to 14 days and was based on the investigator's assessment of a patient's response to therapy according to the criteria listed in table 5.

After baseline examination, the clinical response was evaluated following a score of clinical signs and symptoms (fig. 1) and other laboratory and clinical findings. At the end of treatment, thorax control X ray was carried out.

'Clinical cure' was defined as the subsidence of all clinical signs and symptoms during treatment, 'improvement' as subsidence during treatment but incomplete resolution at the end of treatment, 'failure' as a lack of clinical response during the period of intravenous treatment.

For assessment of the bacterial etiology of disease, appropriate cultures from sputum, tracheal secretions or blood cultures had to be obtained prior to treatment.

Species identification and susceptibility testing of ciprofloxacin and fleroxacin by agar disc diffusion tests were carried out using standard laboratory methods.

MICs were determined by means of the epsilometer test [8] (AB Biodisk, Solna, Sweden) using Mueller-Hinton agar or, if necessary, Mueller-Hinton agar supplemented with 5% sheep blood and *Haemophilus* test medium (Becton Dickinson, Heidelberg, Germany).

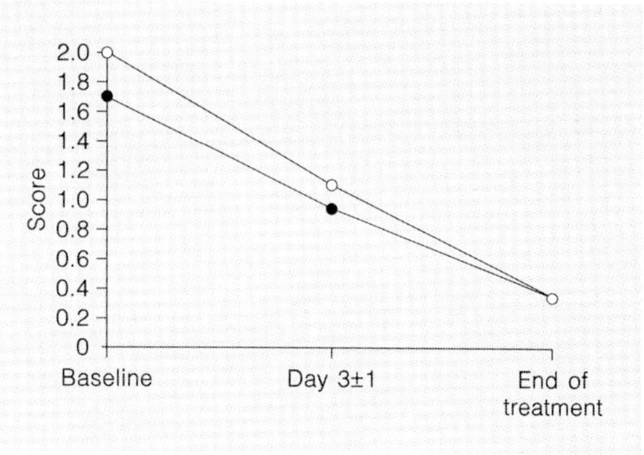

Fig. 1. Mean score of clinical symptoms (dyspnea, cough and rales). The scoring system was as follows: 0 = none; 1 = minor; 2 = moderate; 3 = severe. ● = Fleroxacin; ○ = ciprofloxacin.

Results

The aim of this paper is to present interim results of the ongoing study. To date, 49 patients have been enrolled in the fleroxacin and 53 patients in the ciprofloxacin group. Their demographic data are given in table 6. In both groups, male patients (about 70%) outnumber female patients; the mean age of patients in both groups (55 years) is relatively high. The rate of nosocomially acquired disease is slightly higher in the fleroxacin (61%) than in the ciprofloxacin (55%) group.

Among the bacterial strains isolated at baseline, members of the Enterobacteriaceae (e.g. *E. coli, Klebsiella* spp., *Enterobacter cloacae* and *Proteus mirabilis*) were the predominant infectious agents found in each study group. The MIC values were very low as were those for *H. influenzae* (table 7).

Among the Pseudomonadaceae (e.g. *P. aeruginosa,* and *Stenotrophomonas maltophilia*), two resistant strains were found in each study group. Among the isolates of *S. pneumoniae* which, in general, were only intermediately susceptible to fleroxacin, one resistant

Table 4. Inclusion criteria

Patients ≥18 years old (giving informed consent) with
- nosocomial pneumonia or
- community-acquired pneumonia due to gram-negative bacteria with new and persistent infiltration confirmed by X ray

At least two of the following criteria must be fulfilled:
(1) purulent tracheal secretion (>25 leukocytes/microscopic field, 100:1)
(2) >10,000 leukocytes/μl blood
(3) body temperature >38.5°C
(4) typical auscultatory findings

Table 5. Primary efficacy parameters

Clinical signs and symptoms
- Improvement of dyspnea, cough and rales
- Improvement of X-ray findings
- Body temperature ≤37.5°C at the end of treatment
- Improvement of secretion purulence
- Lack of necessity to change the antibiotic treatment

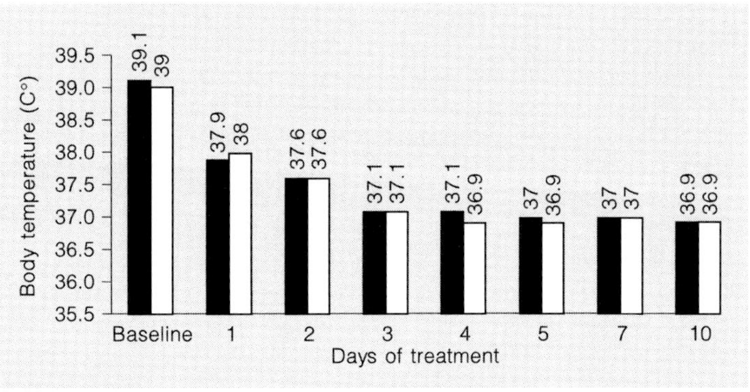

Fig. 2. Course of mean body temperature. ■ = Fleroxacin; □ = ciprofloxacin.

strain was found. All isolates of β-hemolytic streptococci and *S. aureus* were susceptible to both study drugs.

The improvement of clinical symptoms (dyspnea, cough, and rales), which was assessed by means of a scoring system, showed similar results in both treatment groups (fig. 1).

In both study groups, the rapid rate of defervescence was particularly impressive (fig. 2). In patients with a body temperature ≥38.5°C at baseline, 3 days of quinolone treatment resulted in normalization of mean body temperature.

Among the evaluable patients (intent to treat), the following clinical outcome was documented (table 8): at the end of quinolone treatment, cure and improvement rates were 66.7 and 6.7% in the fleroxacin and 70.8 and 8.3% in the comparative group, respectively. The mean duration of quinolone therapy of about 10 days was similar in both groups.

The reasons why 4 patients in the fleroxacin and 5 in the ciprofloxacin group could not be evaluated are listed in table 8.

Twelve patients in the fleroxacin group and 10 in the ciprofloxacin group were recorded as treatment failures. The main reasons for this classification are given in table 9. In each group, lack of radiographic improvement of lung findings, as well as a change in the antimicrobial regimen in the fleroxacin and lack of defervescence in the ciprofloxacin treatment group were the main factors. In those cases where the results of baseline cultures were known, treatment failures were not primarily due to isolates resistant against fleroxacin or ciprofloxacin.

The analysis of adverse events is compiled in table 10. In total, in the fleroxacin group, 17 adverse events were reported in 12 patients (25%), while in the ciprofloxacin group, 10 patients (19%) were affected by 16 adverse events. Among these, in the fleroxacin group, abnormal laboratory values were mainly recorded, while in the ciprofloxacin group, ad-

Table 6. Demographic data

	Fleroxacin (n = 49)	Ciprofloxacin (n = 53)
Gender (male/female)	36/13	36/17
Age[a], years	54.3 ± 15.8	54.8 ± 13.8
Height[a], cm	171.9 ± 8.2	171.6 ± 8.3
Weight[a], kg	72.7 ± 12.9	70.7 ± 14.6
Hospital/community acquired, %	61/39	55/45

[a] Mean ± SD.

Table 7. Bacteriological results at baseline

Isolates	Fleroxacin		Ciprofloxacin	
	total number of isolates[a]	MIC range mg/l	total number of isolates[a]	MIC range mg/l
Enterobacteriaceae	11	0.016–0.25	18	0.012–0.064
Pseudomonadaceae	6 (2)	0.75–32.0	4 (2)	0.094–8.0
Haemophilus influenzae	6	0.03–0.094	10	0.008–0.023
Moraxella catarrhalis	0		1	0.25
Streptococcus pneumoniae	4 (1)	4.0–8.0	5	0.5–2.0
Streptococcus spp.	2	1.0–2.0	1	0.064
Staphylococcus aureus	9	0.125–1.0	5	0.25–1.0

[a] The values in parentheses are the number of resistant strains.

Table 8. Clinical outcome

	Fleroxacin (n = 45)		Ciprofloxacin (n = 48)	
	n	%	n	%
Evaluable patients (intent to treat)				
Cured	30	66.7	34	70.8
Improved	3	6.7	4	8.3
Failure	12	26.6	10	20.8
Days of quinolone therapy	10.0 ± 3.5		10.1 ± 3.6	
Unevaluable patients	4		5	
Pneumococci at baseline	0		1	
Refusal of X ray	1		0	
Dropout due to adverse events	2		2	
Death	1		1	
Missing data (visit 3)	0		1	

verse events related to the digestive (7.5%) and central nervous system (7.5%) were predominantly reported.

Overall, dropouts due to adverse events or concomitant illness were reported in 4 (fleroxacin) and 3 (ciprofloxacin) cases. Deaths (unrelated or remote) were due to cardiac failure (fleroxacin: 2, ciprofloxacin: 1) or to respiratory insufficiency (ciprofloxacin: 2) (table 10).

According to the study protocol, a comprehensive statistical analysis of the clinical and bacteriological response will not be provided before the end of the study, so these data are not yet available.

Table 9. Analysis of treatment failures

	Fleroxacin	Ciprofloxacin
Total failures (n)	12	10
Main reason for classification as failure		
No improvement in X-ray findings	4	4
No improvement in clinical signs	2	2
No defervescence	2	3
Change of antibiotic regimen	4	1
Microbiological results at baseline		
Staphylococci	1	1
Pneumococci	1	2
Gram-negative bacteria	1	2
Gram-positive + gram-negative bacteria	2	1
No positive culture	7	4

Table 10. Adverse events

	Fleroxacin (n = 49)		Ciprofloxacin (n = 53)	
	n	%	n	%
Adverse events				
Digestive system	2	4.1	4	7.5
Central nervous system	1	2.0	4	7.5
Skin and appendages	0		2	3.8
Respiratory system	3	6.1	1	1.9
Cardiovascular system	2	4.1	1	1.9
Other clinical adverse events	3	6.1	1	1.9
Abnormal laboratory values	6	12.2	3	5.7
Adverse events, total	17		16	
Number of patients with adverse events	12	24.5	10	18.9
Dropouts (regardless of causality, excluding deaths)	4		3	
Due to adverse events	2		3	
Due to concomitant illness	2		0	
Deaths (unrelated or remote)	2		3	
Due to cardiac failure	2		1	
Due to respiratory insufficiency	0		2	

Comment

Particularly in cases of nosocomial infection and when elderly patients with risk factors are affected [9], pneumonia is a condition associated with the highest mortality rates (20–50%) of all hospital infections [10].

Several quinolones have been studied for the treatment of nosocomial pneumonia. In general, cure rates with these agents have been comparable with those of more traditional regimens. In a double-blind randomized study, the efficacy of ciprofloxacin was compared to that of imipenem in the treatment of 205 patients with severe pneumonia, primarily associated with intensive care [11]. The clinical and bacteriological efficacy rates were significantly greater for patients treated with ciprofloxacin than for those treated with imipenem.

In a prospective multicenter study published by Lode et al. [2], the cure and improvement rates of ciprofloxacin in a treatment group of patients with lower respiratory tract infections were 39 and 50%, respectively. These data correspond well with the results of patients treated with β-lactams (37% cure and 48% improvement) [2]. In our study, in the ciprofloxacin group, cure or improvement was seen in 38 of a total of 48 patients after intravenous/oral switch and twice daily application of the drug. A similar clinical cure and improvement rate was reported with an intravenous/oral switch regimen of 400 mg fleroxacin once daily (33 of a total of 45 patients).

These interim results of our ongoing study support the conclusion that fleroxacin offers the possibility for once daily dosage and an early switch from parenteral to oral treatment.

Appendix: Participants of the Fleroxacin Study Group

G. Siemon, Donaustauf; A. Balogh, Halle; T. Höfs, Magdeburg; D. Barckow, Berlin; W.H. Bell, Bonn; H. Muschweck, Altdorf; H. Neef, Halle; P.L. Bölcskei, Nürnberg; G. Liebetrau, Lostau; A.G. Mannes, München; A. Sciersky, Burgwedel; H. Worth, Fürth; K. Hahn, Altötting; J. Rockstroh, Bonn; P. Czygan, Neuss; R. Huber, München; P. Leonardt, Leipzig; C. Witt, Berlin; R. Düsing, Bonn; G. Habich, Kutzenberg.

References

1 Scully BE: Therapy of respiratory tract infections with quinolone antimicrobial agents; in Hooper DC, Wolfson JS (eds): Quinolone Antimicrobial Agents. Washington, American Society for Microbiology, 1993, pp 339–362.

2 Lode H, Schaberg T, Höffken G, Borner K: Use of fluoroquinolones in lower respiratory tract infections (LRTI). Int J Antimicrob Agents 1994;4(suppl 2):S47–S52.

3 Cullmann W, Geddes A, Weidekamm E, Urwyler H, Braunsteiner A: Fleroxacin: A review of its chemistry, microbiology, toxicology, pharmacokinetics, clinical efficacy and safety. Int J Antimicrob Agents 1993;2:203–230.

4 Chodosh S: Efficacy of fleroxacin versus amoxicillin in acute exacerbations of chronic bronchitis. Am J Med 1993;94(suppl 3A):131S–135S.

5 Ulmer W: Fleroxacin versus amoxicillin in the treatment of acute exacerbations of chronic bronchitis. Am J Med 1993;94(suppl 3A):136S–141S.

6 Farkas SA: Intravenous fleroxacin versus ceftazidime in the treatment of acute nonpneumococcal lower respiratory tract infections. Am J Med 1993;94(suppl 3A):142S–149S.

7 Fink MP, Snydman DR, Niederman MS, Leeper J, Johnson RH, Wunderink RG, Caldwell JW, Schentag JJ, Siami GA, Zameck RL, Haverstock DC, Reinhart HH, Echols RM, and the Severe Pneumonia Study Group: Treatment of severe pneumonia in hospitalized patients: Results of a multicenter, randomized, double-blind trial. Antimicrob Agents Chemother 1994; 38:547–557.
8 Bolmström A, Arvidson S, Ericsson M, Karlsson A: A novel technique of direct quantification of antimicrobial susceptibility of microorganisms. 26th ICAAC, Los Angeles, 1988.
9 Gómez J, Esquinas A, Agudo MD, Nieto JMS, Núñez ML, Baños V, Canteras M, Valdes M: Retrospective analysis of risk-factors and prognosis in non-ventilated patients with nosocomial pneumonia. Eur J Clin Microbiol Infect Dis 1995;14: 176–181.
10 Penninton JE: Nosocomial pneumonia. Curr Opin Infect Dis 1992;5: 505–511.
11 Bolmström A, Arvidson S, Ericsson M, Karlsson A: A novel technique for direct quantification of antimicrobial susceptibility of microorganisms. 26th ICAAC, Los Angeles, 1988.

Chemotherapy

Chemotherapy 1996;42(suppl 1):43–53

W. Graninger
K. Zedtwitz-Liebenstein
H. Laferl
H. Burgmann

Department of Infectious Diseases and Chemotherapy, University of Vienna, and
Department of Internal Medicine, Kaiser Franz Josef Hospital, Vienna

Quinolones in Gastrointestinal Infections

Key Words
Quinolones
Traveler's diarrhea
Cholera
Typhoid fever

Abstract
Fluoroquinolones are efficient antimicrobial drugs for the treatment of enteric fever, shigellosis, *Escherichia coli* diarrhea, cholera, and traveler's diarrhea. They also play a role in the therapy of yersiniosis, campylobacteriosis, and intestinal salmonellosis. A single dose of quinolones has been effective in the treatment of traveler's diarrhea and cholera. Uncomplicated typhoid fever was cured by norfloxacin, pefloxacin, and ofloxacin 400 mg twice daily (b.i.d.) for 7–14 days or ciprofloxacin 500 mg b.i.d. for 10 days. A single daily dose of 400 mg fleroxacin for 3 days has been shown to be effective in this indication. A few reports suggest that the newer quinolones can eliminate the carrier state. The efficacy of quinolones in the prophylaxis and treatment of intra-abdominal infections following abdominal surgery requires further investigation.

Quinolones show favorable activity against Enterobacteriaceae and other gram-negative aerobes. They have little activity against gram-positive cocci and anaerobic rods. They reach high concentrations in the gut wall and stools. Quinolones have demonstrated their efficacy in the treatment of gastrointestinal infections, notably in acute bacterial diarrhea, cholera, typhoid fever, and intra-abdominal surgery. Some clinically important aspects of this therapeutic area are summarized in this paper.

KARGER
E-Mail karger@karger.ch
Fax +41 61 306 12 34

0009–3157/96/0427–0043$10.00/0

Wolfgang Graninger, MD
Department of Infectious Diseases and Chemotherapy
University Clinic of Internal Medicine I
Währinger Gürtel 18–20
A–1090 Vienna (Austria)

Acute Bacterial Diarrhea

Diarrhea is a frequent medical problem. Various infectious agents may cause diarrhea, but bacteria are responsible for 80% of cases. The pathogens responsible include *Escherichia coli, Vibrio cholerae, Vibrio parahaemolyticus, Shigella* spp., *Salmonella* spp., *Plesiomonas* spp., and *Aeromonas* spp. Salmonellosis has been associated with classic enteric fever or severe diarrhea with abdominal pain, whereas *Shigella* enteritis typically has an abrupt onset with bloody or watery stools, abdominal pain, and fever. Enteritis associated with enterotoxigenic *E. coli* varies considerably but is usually of relatively short duration [1]. *Campylobacter* enteritis has also been described as intense diarrhea with abdominal pain and fever in more than 50% of cases with normal immunity. Patients with diarrhea should supplement the therapy by taking fluid and electrolytes. Drugs which act by alleviating the symptoms, such as bismuth salicylate and loperamide, are effective to a certain degree. Bacterial gastroenteritis is often a self-limited disease; however, in a significant number of people, the illness continues beyond 5–10 days or complications develop. Antimicrobial agents are the mainstay of therapy, the fluoroquinolones being the most active against all pathogens involved [2, 3]. Antibiotics might be indicated in febrile patients with complicating factors.

In patients with shigellosis, a single dose of norfloxacin 800 mg was found to be as effective as 5-day treatment with co-trimoxazole 160/800 mg twice daily (b.i.d.) [4]. In another study, 21 patients with shigellosis were treated with ofloxacin 400 mg in a single dose or ofloxacin 200 mg three times a day (t.i.d.) for 1 day [5]. Rates of clinical cure and bacteriological eradication were identical in both groups. A single dose of 1 g ciprofloxacin was found to be an effective therapy for patients infected with species of *Shigella* other than *Shigella dysenteriae* type 1. In this study, single-dose therapy was found to be inferior to treatment with 10 doses for patients with *S. dysenteriae* type 1 [6].

In a double-blind, randomized trial, 85 patients with acute bacterial diarrhea received either 500 mg ciprofloxacin twice daily or placebo. The main causative organisms were *Campylobacter jejuni, Salmonella* spp., and *Shigella*. The mean duration of fever and diarrhea was significantly reduced in patients treated with ciprofloxacin compared with patients in the placebo group [7].

A group of 508 adults with acute diarrhea treated with fleroxacin 400 mg given orally as a single dose or once daily for 3 days showed a more rapid clinical improvement than placebo-treated patients. Pathogens included *V. cholerae, V. parahaemolyticus, Shigella* spp., *Salmonella* spp., *Plesiomonas* spp., *Aeromonas* spp., and *E. coli*. Stool cultures on days 3 and 5 were negative in 94% of patients treated with single doses of fleroxacin and in 93% of those treated with three doses of fleroxacin, as compared with 57% of patients treated with placebo. Comparing placebo with fleroxacin in the treatment of acute bacterial diarrhea confirms the conclusion that single-dose therapy should be advantageous for the empirical treatment of certain diarrheal diseases [8].

Quinolones are also the drugs of choice for traveler's diarrhea. Ciprofloxacin in a single, 500-mg dose was compared to placebo in the treatment of acute traveler's diarrhea in British soldiers in Belize, with 45 patients receiving ciprofloxacin and 38 placebo. A single, 500-mg dose of ciprofloxacin was found to be an effective empirical treatment for reducing the duration and severity of traveler's diarrhea [9]. A recently reported double-blind, placebo-controlled study of fleroxacin in 195 patients with traveler's diarrhea demon-

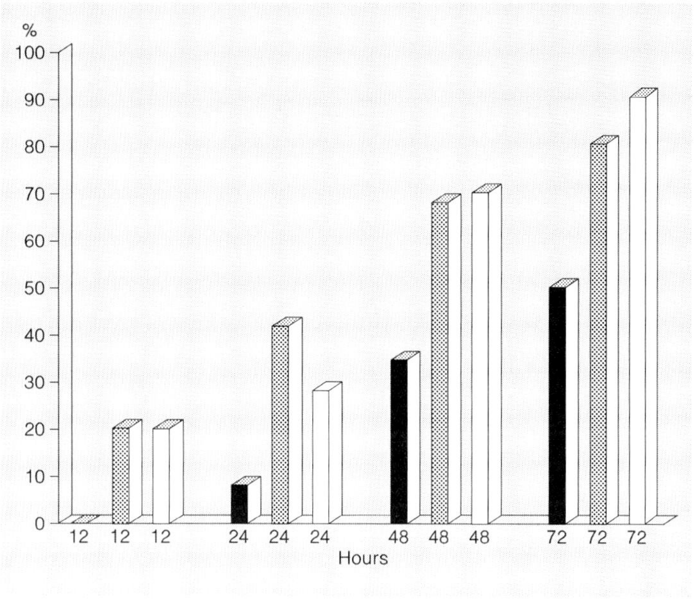

Fig. 1. Proportion of patients cured of traveler's diarrhea. ■ = Placebo; ▩ = 1 × 400 mg fleroxacin; □ = 2 × 400 mg fleroxacin.

strated that a single dose of 400 mg fleroxacin is effective [10] (fig. 1). The single-dose treatment was as effective as the 2-day regimen (400 mg once daily), with cure rates of 80 and 81%, respectively. The mean time to production of the last unformed stool was 27 h in the two fleroxacin groups and 45 h in the placebo group. Adverse events, particularly minor neuropsychiatric disturbances such as headache and insomnia, were more common among patients receiving the active treatment.

Enteric Infections in Children

Epidemic infantile diarrhea has long been recognized as a potentially serious problem in newborn nurseries. Its mortality rate has been as high as 24–50% [11, 12]. Epidemic diarrhea among hospitalized newborns has been associated with certain enteropathogenic serotypes of *E. coli* (EPEC). However, in many areas such as South Africa and southern Brazil, EPEC remain the most common cause of diarrhea in infants and young children especially during the summer months [13–15]. Endemic childhood diarrheal illness in areas like England and Canada also remains associated with EPEC in 6–18% of cases [13, 14]. Another cause of enteric infection in children is shigellosis.

Drugs of choice for infantile diarrhea are β-lactams or trimethoprim-sulfamethoxazole (TMP-SMZ). Nalidixic acid has most commonly been used in children to treat shigellosis due to strains that are resistant to ampicillin and TMP-SMZ [16–20]. The clinical cure rate associated with nalidixic acid has been found to be comparable to that achieved with ampicillin, although eradication of *Shigella* from the stools may take longer [21]. These findings, however, contradict those of another controlled clinical trial of nalidixic acid to treat shigellosis in children [22]. Resistance of *Shigella* to nalidixic acid has already been reported [23, 24] and it can be expected that in communities where nalidixic acid is commonly used, widespread resistance will quickly develop. Resistance to ampicillin, TMP/SMX, and nalidixic acid raises the question of whether newer quinolones with their favorable resistance pattern might be used in children.

Sixteen children with severe salmonellosis were treated with ampicillin, amoxicillin or co-trimoxazole, but 7 experienced clinical failure after 4–8 days' therapy. The severe bloody diarrhea persisted and stool cultures remained positive in all cases. These children were then given 12 mg/kg/day of pefloxacin orally in 2 daily doses for 7 days. No exanthema, nausea, joint pain, or other side effects were seen. Stool cultures became negative in 5 cases. In 1 patient, a relapse of enteric fever was observed and another became an asymptomatic carrier of *Salmonella enteritidis* without clinical relapse [25].

The main side effect of quinolones in children may be the toxicity for immature joint cartilage, as described in some animal models. Nalidixic acid is approved by the US Food and Drug Administration for use in children over 2 months of age, while the newer quinolones are not yet approved for pediatric use. Nalidixic acid causes joint disease when given in high doses to immature animals of a number of species [26]. In children with cystic fibrosis treated with ciprofloxacin for 3 months, no evidence of joint disease was found [27, 28]. Joint pains or stiffness are sometimes described with high doses of fluoroquinolones [28, 29]. There is only one case report of destructive polyarthropathy in a 17-year-old boy following treatment with pefloxacin 800 mg/day for 3 months, but an underlying joint disease could not be excluded [30]. While pediatricians are often fond of prescribing nalidixic acid, it must be kept in mind that it is also a quinolone. Questions about the use of quinolones in children are not settled yet; however, it is agreed that in life-threatening situations they can be used readily.

Cholera

Cholera is endemic in parts of Asia, the Middle East, Africa and South America, and is also present in the Ukraine, Romania, and Albania, whence it may be transported to western European countries. This should be borne in mind when patients present with profuse watery diarrhea, vomiting, muscle cramps, and dehydration. Treatment of cholera requires rapid correction of hypovolemia and metabolic acidosis and prevention of hypokalemia. Antimicrobial treatment commonly comprises tetracycline, or TMP-SMZ. Multiple antibiotic resistance now seems to be a growing problem [31].

In one study, 202 cholera patients were treated with ciprofloxacin 250 mg once daily for 3 days or with tetracycline 500 mg 4 times a day (q.i.d.) for 3 days. The clinical and bacteriologic efficacy rates were similar in both groups and the results showed that ciprofloxacin given once daily was as effective as the standard tetracycline regimen [32]. It has recently been shown that fleroxacin as a single oral dose of 400 mg is as effective as the standard 3-day regimen of tetracycline (500 mg, every 6 h) in eradicating *V. cholerae* and effecting a clinical cure.

Prophylaxis with antimicrobial agents has been discussed. Since cholera is spread by ingestion of contaminated water or food, prophylaxis includes the same measures as for traveler's diarrhea. In almost all of the affected populations, risk factors for cholera and other enteric diseases have included having a family member with diarrhea, lack of proper sanitation, and other conditions associated with poverty [33, 34]. Cholera vaccines have also proved to be ineffective in areas where the disease is endemic [35]. Chemotherapeutic prophylaxis (with tetracycline) is recommended only to limit secondary cases among domestic contacts. Compared with other agents for the prophylaxis of cholera, the fluoroquinolones appear to be the best choice. These drugs produce high concentrations in the feces for several days and the development of resistance is relatively rare [36]. Chemoprophylaxis, however, has never stopped the transmission of cholera. A randomized, double-blind study in Lima, Peru, examined the tolerability and efficacy of a single 250-mg dose of ciprofloxacin in preventing diarrhea and *V. cholerae* O1 infection among household contacts. No reduction in either the occurrence of diarrhea or the acquisition of *V. cholerae* among the household contacts was found. The benefit observed was for those household contacts who were already infected at the time of enrollment in the study. In this subgroup, a significant reduction of the bacterial load was noted [37].

Typhoid Fever

Another indication for the use of quinolones is typhoid fever, which continues to be a common endemic disease in developing countries, constituting a risk for travelers [38]. Transmission is chiefly via food that has been contaminated with strains of *Salmonella typhi* by healthy carriers during its preparation. Traditionally, the drugs of choice have been chloramphenicol, ampicillin, or TMP-SMZ. However, in recent years, *S. typhi* has developed resistance to these agents, mediated by R plasmid transfer [39]. An effective alternative for the treatment of typhoid fever is represented by the third-generation cephalosporins. For example, with ceftriaxone, a cure rate of 91% has been achieved [40]. Fluoroquinolones also possess a high degree of in vitro activity against *Salmonella* species, including multiresistant strains, and comparative clinical studies have shown their efficacy to be comparable or superior to standard antityphoid therapy [41–43]. Moreover, fluoroquinolones are more effective than traditional antityphoid agents in eradicating intestinal and biliary reservoirs of infection in the chronic carrier state [42], and have minimal effect on the intestinal anaerobic flora. The penetration of quinolones into macrophages seems to be important for their effectiveness in systemic *Salmonella* infections.

Early studies with norfloxacin 400 mg b.i.d. for 14 days demonstrated a cure rate of 92% in 12 patients with uncomplicated typhoid fever [44]. A similar cure rate was achieved with pefloxacin 400 mg b.i.d. for 7 days [45]. Ciprofloxacin was effective in the treatment of 38 patients with typhoid fever confirmed by blood culture. The drug was administered in a dose of 500 mg every 12 h for 14 days. There was one treatment failure in a patient developing acute cholecystitis [46]. The same cure rate of 97% was achieved in 35 patients given ofloxacin 400 mg b.i.d. for 7–10 days [47]. Ofloxacin 300 mg b.i.d. for 10–14 days in 64 patients with typhoid fever revealed a clinical cure rate of 100% [48]. In two comparative studies, oral fleroxacin 400 mg once daily for 7–14 days was at least as effective as chloramphenicol 50 mg/kg/day for 14 days [49, 50]. No relapse occurred in those who received 10 or 14 days' treatment with fleroxacin, but was documented in 17% of those who received the 7-day fleroxacin regimen and in 6% of those who were treated with chloramphenicol [49, 50]. In a recent Vietnamese study, 3 days of fleroxacin yielded the same results in typhoid fever as a 5-day course [Duong et al., pers. commun.], with all patients being cured. Reasons for these excellent results with the shorter duration treatment may be that the Vietnamese strains are particularly susceptible, or that the lower mean body weight in this country leads to higher drug concentrations when using a fixed dosing schedule. Thus, the minimum duration of typhoid therapy still needs to be clarified.

Norfloxacin, ciprofloxacin, or ofloxacin treatment resulted in a cure in 80–90% of chronic enteric carriers of *S. typhi*. Chronic typhoid carriers were treated with either 400 mg norfloxacin or placebo every 12 h for 28 days. Of 12 patients treated with norfloxacin, 11 exhibited negative stool and bile cultures at the end of therapy, while all 12 placebo-treated subjects remained positive [51]. In another study, ciprofloxacin 750 mg was administered twice daily for 28 days to 12 typhoid carriers. The bacteriological cure rate was 92% (11 of 12 patients) [52].

Prophylaxis of Traveler's Diarrhea

Diarrhea is by far the most common medical problem among people traveling to semitropical and tropical areas of the developing world, affecting 20–50% [53]. Onset is usually between 3 and 15 days after arrival. The most common sources of traveler's diarrhea are contaminated food [54] and, to a lesser degree, contaminated water. A more detailed review of the prevention and treatment of traveler's diarrhea has recently been published by DuPont and Ericsson [55]. At least 80% of traveler's diarrhea cases are caused by bacterial enteropathogens [55], which explains the remarkable prophylactic and therapeutic effect of antimicrobial drugs. Overall, the most important etiologic agents of traveler's diarrhea are enterotoxigenic *E. coli* (up to 50%), *Shigella* spp., *C. jejuni*, *Aeromonas* spp., *Plesiomonas shigelloides*, *Salmonella* spp., and noncholera vibrios [55]. There are seasonal and regional differences in the occurrence of the enteropathogens. The best prophylaxis is, of course, avoiding contaminated food and water. Many travelers, however, prefer to take antimicrobial drugs and bismuth subsalicylate for the prevention of diarrhea, which have been acknowledged as effective by the National Institutes of Health [56]. Prophylactic antibacterial drugs eliminate approximately 90% of the cases of traveler's diarrhea that would occur without their use [57]. Nevertheless, we do not recommend prophylaxis with these preparations because traveler's diarrhea is a self-limited disorder, and because of the risk of side effects or the emergence of resistance [58]. However, chemoprophylaxis does have a place in preventing illness in some travelers, as long as certain relevant factors are taken into consideration. These include underlying health, the importance of the trip, the traveler's wishes concerning prophylaxis, and the traveler's willingness to restrict the diet [59]. If a decision is taken in favor of prophylactic drugs, the following are recommended at present, depending on the geographic region: fluoroquinolones, TMP-SMZ, or doxycycline. Prophylaxis should not be instituted if the period of the risk exceeds 3 weeks [41]. A study showed that only 7% of travelers receiving norfloxacin 400 mg once daily for 14 days (n = 56) developed diarrhea, compared with 61% of patients receiving placebo (n = 59) [60]. Prophylaxis (during the stay in a high-risk area) with 400 mg norfloxacin (n = 105) once daily offered a 93% protection rate compared to 74% with placebo (n = 117) [61], while 500 mg ciprofloxacin once daily achieved a 94% protection rate compared to 36% with placebo in travelers to Tunisia (n = 53) [62].

Mesenteric Adenitis

This illness, which closely mimics appendicitis, is caused by *Yersinia enterocolitica* and *Yersinia pseudotuberculosis*. It is usually a self-limited disease; however, septicemia with a high mortality rate may develop. Again, quinolones are the drugs of choice.

A typical complication of infection due to *Yersinia* spp. is chronic arthritis. Reactive arthritis [63] is a sterile joint disease following a primary infection elsewhere in the body. *Yersinia*, *Salmonella*, *Campylobacter* and *Chlamydia* species have been implicated as causal pathogens. The effect of a long course of ciprofloxacin (500 mg orally twice daily for 3 months) in reactive arthritis was assessed in a double-blind prospective trial. A total of 36 patients with arthritis of 2 months to 16 years' duration were included in the study: 19 received ciprofloxacin and 17 placebo. The triggering infectious agents had been *Chlamydia* (4 patients), *Campylobacter* (1 patient), and *Yersinia* (31 patients). Ciprofloxacin treatment had a favorable effect on joint symptoms as indicated by the scores for arthralgia, pain at movement, difficulty in moving around, and morning stiffness as reported by the patients.

Whipple's Disease

No studies on the effect of quinolones in this disease have been published. TMP-SMZ is still the drug of choice. Alternatives include ceftriaxone and chloramphenicol.

Tropical Sprue

No studies with quinolones have been performed. Tetracyclines are still the drugs of choice.

Helicobacter pylori

The newer quinolones inhibit growth of *Helicobacter pylori*, which has been associated in the pathogenesis of antral gastritis. The quinolones must be able to penetrate the gastric mucosa in which the organisms survive. Recurrence and the emergence of resistance have been reported after treatment with ofloxacin 200 mg b.i.d. for 21 days. Eighteen patients were treated, of whom only 13 were evaluable and only one had *H. pylori* eradicated from the antrum. Resistance to ofloxacin emerged during treatment [64]. These observations suggest that the newer quinolones may not be effective in the treatment of *H. pylori*-associated gastritis.

Peritonitis and Intra-Abdominal Infections

The quinolones are active against nearly all of the aerobic bacterial pathogens that commonly produce peritonitis, cholangitis, and biliary sepsis. Quinolones would be a good choice in the treatment of postsurgical intra-abdominal infections because of their excellent tissue penetration and broad antibacterial spectrum. Little data on the role of quinolones in the treatment of surgical infections are available; however, on the basis of the few studies carried out to date, it can be concluded that surgical infections due to enterobacteria may be successfully treated with the new quinolones. When streptococci, enterococci or anaerobic bacteria are involved, antimicrobial agents effective against these microorganisms must be added. The data suggest that generally the quinolones are as efficacious as the other antibiotics [65].

The quinolones appear to be effective in most patients with peritonitis secondary to chronic ambulatory peritoneal dialysis

(CAPD). In patients with peritonitis associated with CAPD, ciprofloxacin in a dose of 25 or 50 mg/l of dialysate for 5–7 days was shown to be an effective therapy [66, 67]. Ciprofloxacin at 20 mg/l added to each dialysate bag for 10 days was as effective as standard intraperitoneal vancomycin and gentamicin [68]. Oral ciprofloxacin for 10 days (250–500 mg at the time of each dialysate exchange) was as effective as intraperitoneal vancomycin plus netilmicin in CAPD peritonitis [69]. Failures with *Staphylococcus epidermidis* have occurred, however.

Spontaneous bacterial peritonitis (SBP) is defined as bacterial peritonitis with ascites. It is noted primarily in patients with cirrhosis and hepatic functional impairment. More than 92% of all cases of SBP are monomicrobial, with aerobic gram-negative bacilli, *E. coli* (nearly 50%), and *Klebsiella* species being responsible for more than two thirds of all cases. Gram-positive organisms are implicated in 25% of all cases, with staphylococcal and streptococcal species being the most common. Most studies show a mortality rate of about 30–40%. Secondary prophylaxis against recurrent SBP was effective in one study in which 400 mg of norfloxacin/day decreased the 1-year probability of recurrence from 68 to 20% [70].

Prophylaxis in Abdominal Surgery

Several studies of colorectal surgery have shown that preoperative antibiotic use reduces the rate of postoperative infection (particularly with *E. coli* among the aerobes and *Bacteroides fragilis* among the anaerobes).

In a study comparing oral or intravenous pefloxacin (n = 46) at high and low doses with low-dose cefuroxime (n = 46) in patients undergoing colorectal surgery, a 'success rate' of 91.5 and 84.4%, respectively, was seen [71]. In a double-blind trial, a single intravenous dose of ciprofloxacin 200 mg and a single intravenous dose of latamoxef 2 g were compared [72]. A single dose of intravenous metronidazole 500 mg was added to each regimen for anaerobic coverage. A postoperative wound infection rate of less than 10% was achieved in both the ciprofloxacin and latamoxef groups. In another study, prophylaxis in colorectal surgery with 80 mg gentamicin plus 500 mg metronidazole (718 patients) was compared to 750 mg ciprofloxacin plus 500 mg metronidazole (100 patients). The wound infection rate of 3% in the ciprofloxacin group was significantly lower than the 13.4% obtained with the aminoglycoside prophylaxis [73].

Biliary tract surgery does not usually require prophylaxis. However, in patients with risk factors such as age over 70 years, a past history of biliary tract surgery, obstructive jaundice, or choledocholithiasis [74–77], antibiotic prophylaxis is required because of the high rate of postoperative wound infection (up to 25%). The concentration of the fluoroquinolones [78] in the biliary tract is high enough to prevent infections.

Cooreman et al. [79] assessed the efficacy of pefloxacin and cefazolin given as single intravenous doses of 800 mg or 2 g, respectively, before biliary tract surgery (cholecystectomy in high-risk patients or accompanied by common bile duct exploration). One polymicrobial wound abscess was noted in the pefloxacin group (n = 53; *E. coli, S. epidermidis*, viridans streptococci, *S. millieri*) and 1 intra-abdominal abscess *(Staphylococcus aureus)* and 1 case of peritonitis *(Klebsiella pneumoniae, Pseudomonas aeruginosa)* in the cefazolin group (n = 50) [79]. In another study, a single oral dose of ciprofloxacin 200 mg was compared with a single dose of ceftriaxone 2 g administered preoperatively in 200 patients undergoing biliary surgery

[80]. The bacteria isolated corresponded to the typical intestinal flora known to be present in the contaminated biliary tract [81–83]. The most frequent operations performed were cholecystectomy, choledochal revisions, and transduodenal papillotomy. No difference between the two groups was detected in the prevention of postoperative wound infection. Ciprofloxacin 750 mg orally twice daily was as effective as cefazolin 1 g twice daily in 116 patients undergoing endoscopic retrograde cholangiopancreatography [84].

References

1 Mattila L: Clinical features and duration of traveler's diarrhea in relation to its etiology. Clin Infect Dis 1994;19:728–734.

2 Ericsson CD, Johnson PC, Du Pont HL, Morgan DR, Bitsura JAM, de la Cabada FJ: Ciprofloxacin or trimethoprim-sulfamethoxazole as initial therapy for traveler's diarrhea: A placebo-controlled, randomized trial. Ann Intern Med 1987;106: 216–220.

3 Du Pont HL, Ericsson CD, Mathewson JJ, Du Pont MW: Five versus three days of ofloxacin therapy for traveler's diarrhea: A placebo-controlled study. Antimicrob Agents Chemother 1992;36:887–891.

4 Gotuzzo E, Oberhelman RA, Maguina C, Berry SJ, Yi A, Guzman M, Ruiz R, Leon Barua R, Sack RB: Comparison of single-dose treatment with norfloxacin and standard 5-day treatment with TM-SMZ for acute shigellosis in adults. Antimicrob Agents Chemother 1989;33: 101–104.

5 Akalin HE: Antibiotic treatment of acute bacterial diarrheas: The role of new quinolones. Ther Infect Dis 1991;6:205–213.

6 Bennish ML, Salam MA, Khan AM: Treatment of shigellosis. III: Comparison of one- or two-dose ciprofloxacin with standard 5-day therapy. Ann Intern Med 1992;117:727–734.

7 Pichler ET, Diridl G, Stickler K: Clinical efficacy of ciprofloxacin compared with placebo in bacterial diarrhea. Am J Med 1987; 82(suppl 4A):329–332.

8 Butler T, Lolekha S, Rasidi C, Kadio A, del Rosal PL, Iskadar H, Rubinstein E, Pastore G: Treatment of acute bacterial diarrhea: A multicenter international trial comparing placebo with fleroxacin given as a single dose or once daily for 3 days. Am J Med 1993;94(suppl 3A):187–194.

9 Salam I, Katelaris P, Leigh-Smith S: Randomised trial of single-dose ciprofloxacin for traveler's diarrhoea. Lancet 1994;344:1537–1539.

10 Steffen R, Jori J, DuPont HL, Mathewson JJ, Stürchler D: Treatment of traveller's diarrhoea with fleroxacin: A case study. J Antimicrob Chemother 1993;31:767–776.

11 Jacobs S, Holzel A, Wolman B, Keen JH, Miller V, Taylor J, Gross RJ: Outbreak of infantile gastroenteritis caused by *E. coli* 0114. Arch Dis Child 1970;45:656.

12 Neter E: Enteritis due to enteropathogenic *E. coli*: Present day status and unsolved problems. J Pediatr 1959;55:223.

13 Levine MM: Enteropathogenic *E. coli* of classic serotypes associated with infant diarrhea: Epidemiology and pathogenesis. Epidemiol Rev 1984;6:31–51.

14 Gurwith M: A prospective study of enteropathogenic *E. coli* in endemic diarrheal disease. J Infect Dis 1978; 137:292.

15 Toledo MRF: Enteropathogenic *E. coli* serotypes and endemic diarrhoea in infants. Infect Immun 1983;39:586–589.

16 Frost JA, Rowe B: Acquisition of trimethoprim resistance in epidemic strain of *Shigella dysenteriae* type 1 from Zaire. Lancet 1982;i:963.

17 Parry HE: Nalidixic acid for shigellosis. Lancet 1983;ii:1206.

18 Malengreau M: Nalidixic acid in *Shigella dysenteriae* outbreaks. Lancet 1984;ii:172.

19 Bose R, Nashipuri JN, Sen PK: Epidemic of dysentery in West Bengal: Clinicians' enigma. Lancet 1984;ii: 1160.

20 Moorhead PJ, Parry HE: Treatment of Sonne dysentery. Br Med J 1965; ii:913–915.

21 Hardy AV, Cummins SD: Studies of the acute diarrheal diseases: A preliminary note on the clinical response to sulfadiazine therapy. Public Health Rep 1943;58:693–696.

22 Haltalin KC, Nelson JD: Comparative efficacy of nalidixic acid and ampicillin for severe shigellosis. Arch Dis Child 1973;48:305–312.

23 Panhotra BR, Desai B: Nalidixic-acid-resistant *Shigella dysenteriae* 1. Lancet 1985;1:763.

24 Munshi MH, Sack DA: Plasmid-mediated resistance to nalidixic acid in *Shigella dysenteriae* type 1. Lancet 1987;ii:419–421.

25 Gendrel D, Raymond J, Legall M: Use of pefloxacin after failure of initial antibiotic treatment in children with severe salmonellosis. Eur J Clin Microbiol Infect Dis 1993;12:209–211.

26 Ingham B, Brentnall DW: Arthropathy induced by fused N-alkyl-4-pyridone-3-carboxylic acids. Toxicol Lett 1977;1:21–26.

27 Kubin R: Safety and efficacy of ciprofloxacin in paediatric patients. Infection 1993;21:413–421.

28 Schaad UB, Wedgwood-Krucko J: Nalidixic acid in children: Retrospective matched controlled study for cartilage toxicity. Infection 1987; 15:165–168.

29 Schaad UB, Stoupis C, Wedgwood J, Tschaeppeler H: Clinical, radiologic and magnetic resonance monitoring for skeletal toxicity in pediatric patients with cystic fibrosis receiving a 3-month course of ciprofloxacin. Pediatr Infect Dis 1991;10: 723–729.
30 Chevalier X, Albengres E, Voisin MC: A case of destructive polyarthropathy in a 17-year-old youth following pefloxacin treatment. Drug Safety 1992;7:310–314.
31 Glass RI, Becker S, Huq MI, Stoll BJ: Endemic cholera in rural Bangladesh. Am J Epidemiol 1982;116: 959–970.
32 Gotuzzo E, Seas C, Echevarria J, Carillo C: Ciprofloxacin for the treatment of cholera: A randomized, double-blind, controlled clinical trial of a single daily dose in Peruvian adults. Clin Infect Dis 1995;20: 1485–1490.
33 Tamayo JF, Mosley WH, Alvero MG, Joseph PR, Gomez CZ, Montague T, Dizon JJ, Henderson DA: Studies of cholera El Tor in the Philippines. 3: Transmission of infection among household contacts of cholera patients. Bull World Health Organ 1965;33:645–649.
34 Pakistan-Seato Cholera Research Laboratory: Cholera in East Pakistan families: 1962–1963. Bull World Health Organ 1965;32:205–209.
35 Sommer A, Khan M, Mosley WH: Efficacy of vaccination of family contacts of cholera cases. Lancet 1973;i:1230–1232.
36 Crump B, Wise R, Dent J: Pharmacokinetics and tissue penetration of ciprofloxacin. Antimicrob Agents Chemother 1983;24:784–786.
37 Echevarria J, Seas C, Carrillo C: Efficacy and tolerability of ciprofloxacin prophylaxis in adult household contacts of patients with cholera. Clin Infect Dis 1995;20:1480–1484.
38 Edelman R, Levine MM: Summary of an international workshop on typhoid fever. Rev Infect Dis 1986;8: 329–349.
39 Schwalbe RS, Hoge CW, Morris JG, O'Hanlon PN, Crawford RA, Gilligan PH: In vivo selection for transmissible drug resistance in *Salmonella typhi* during antimicrobial therapy. Antimicrob Agents Chemother 1990;34:161–163.
40 Asma I, Thomas B, Samir KN, Nurul HA, Klaus S, Harol BH, Arnold IS: Randomized treatment of patients with typhoid fever by using ceftriaxone, or chloramphenicol. J Infect Dis 1988;158:742–747.
41 Du Pont HL, Ericsson CD, Du Pont MW: Emporiatric enteritis: Lessons learned from U.S. students in Mexico. Trans Am Clin Climatol Assoc 1985;97:32–42.
42 Trujillo Z, Quiroz C, Gutierrez MA, Arias J, Renteria M: Fluoroquinolones in the treatment of typhoid fever and the carrier state. Eur J Clin Microbiol Infect Dis 1991;10:334–341.
43 Balfour JA, Todd PA, Peters DH: Fleroxacin: A review of its pharmacology and therapeutic efficacy in various infections. Drugs 1995;49: 749–850.
44 Velmonte MA, Montalban CS: Norfloxacin in the treatment of infections caused by *Salmonella typhi.* Scand J Infect Dis 1988;56(suppl): 46–48.
45 Ait-Khaled A, Zidane L, Amrane A, Aklil R: A seven-day pefloxacin course for the treatment of typhoid fever in Algeria. Rev Infect Dis 1989;11(suppl 5):S1191–S1192.
46 Ramirez CA, Bran JL, Mejia CR, Garcia JF: Open prospective study of the clinical efficacy of ciprofloxacin. Antimicrob Agents Chemother 1985;28:128–132.
47 Tanphaichitra D, Srimuang S: In vitro and clinical evaluation of ofloxacin in urinary tract infection and enteric fever. Rev Infect Dis 1989; 11(suppl 5):S1190–S1191.
48 Wang F, Gu XJ, Zhang MF,Tai TY: Treatment of typhoid fever with ofloxacin. J Antimicrob Chemother 1989;23:785–788.
49 Arnold K, Hong CS, Nelwan R, Zavala Trujillo I, Kadio A, Barros MA, de Garis S: Randomized comparative study of fleroxacin and chloramphenicol in typhoid fever. Am J Med 1993;94(suppl 3A):195–200.
50 Abejar NH, Dimaano EM, Cabanban AB, Tayag EA: Comparative study of fleroxacin vs chloramphenicol in typhoid fever at San Lazaro Hospital. Philippine J Microb Infect Dis 1993;22:8–10.
51 Gotuzzo EJ, Guerra G, Benavente L, Palomino JC, Carrillo C, Lopera J, Delgado F, Nalin DR, Sabbaj J: Use of norfloxacin to treat chronic typhoid carriers. J Infect Dis 1988; 137:1221–1225.
52 Ferrecio C, Morris JG Jr, Valdivieso C, Prenzel I, Sotomayor V, Drusano GL, Levine MM: Efficacy of ciprofloxacin in the treatment of chronic typhoid carriers. J Infect Dis 1988; 157:1235–1239.
53 Steffen R: Epidemiologic studies of traveler's diarrhea, severe gastrointestinal infections, and cholera. Rev Infect Dis 1986;8(suppl 2):122–130.
54 Wood LV, Ferguson LE, Hogan P, Thurman D, Morgan DR, Du Pont HL, Ericsson CD: Incidence of bacterial enteropathogens in foods from Mexico. Appl Environ Microbiol 1983;46:328–332.
55 Du Pont HL, Ericsson CD: Prevention and treatment of traveler's diarrhea. N Engl J Med 1993;328: 1821–1827.
56 Gorbach SL, Edelman R (eds): Traveler's diarrhea: National Institutes of Health Consensus Development Conference. Rev Infect Dis 1986;8(suppl 2):109–233.
57 Farthing MJG, Du Pont HL, Guandalini S, Keusch GT, Steffen R: Treatment and prevention of traveler's diarrhoea. Gastroenterol Int 1992;5:162–175.
58 Rademaker CM, Hoepelman IM, Wolfhagen MJ, Beumer H, Rozenberg Arska M, Verhoef J: Results of a double-blind placebo-controlled study using ciprofloxacin for prevention of traveler's diarrhea. Eur J Clin Microbiol Infect Dis 1989;8: 690–694.
59 Farrar WE: Antibiotic resistance in developing countries. J Infect Dis 1985;152:1103–1106.
60 Johnson PC, Ericson CD, Morgan DR, Du Pont HL, Cabada FJ: Lack of emergence of resistant fecal flora during successful prophylaxis of traveler's diarrhea with norfloxacin. Antimicrob Agents Chemother 1986;30:671–674.
61 Scott DA, Haberberger RL, Thornton SA, Hyams KG: Norfloxacin for prophylaxis of traveler's diarrhea in U.S. military personnel. Am J Trop Med Hyg 1990;42:160–164.

62 Rademaker CMA, Hoepelman IM, Wolfhagen MJHM, Beumer H, Rozenberg-Arska M, Verhoef J: Results of a double-blind placebo-controlled study using ciprofloxacin for prevention of traveler's diarrhea. Eur J Clin Microbiol Infect Dis 1989;8(8): 690–694.
63 Timo Y, Reijo L: Treatment of chronic reactive arthritis with a 3-month course of ciprofloxacin. Drugs 1993;45(suppl 3):452–453.
64 Glupczynski Y, Labbe M: Treatment failure of ofloxacin in *Campylobacter pylori* infection. Lancet 1987;ii:1096.
65 Nord CE: Surgical prophylaxis and treatment of surgical infections with quinolones. Rev Infect Dis 1989; 11(suppl 5):S1287–S1291.
66 Dryden MS, Wing AJ, Phillips I: Low dose intraperitoneal ciprofloxacin for the treatment of peritonitis in patients receiving continuous ambulatory peritoneal dialysis. Antimicrob Agents Chemother 1991;28: 131–139.
67 Ludlam HA, Barton I, White L, McMullin C: Intraperitoneal ciprofloxacin for the treatment of peritonitis in patients receiving continuous ambulatory peritoneal dialysis (CAPD). J Antimicrob Chemother 1990;25:843–851.
68 Friedland JS, Iveson TJ: A comparison between intraperitoneal ciprofloxacin and intraperitoneal vancomycin and gentamicin in the treatment of peritonitis associated with continuous ambulatory peritoneal dialysis (CAPD). J Antimicrob Chemother 1990;26(suppl F):77–81.
69 Topson JS, Orr KE, George JC, Stansfield E: A comparison between ciprofloxacin and intraperitoneal vancomycin and netilmicin in CAPD peritonitis. J Antimicrob Chemother 1990;26(suppl F):63–71.
70 Gilbert JA, Kamath PS: Spontaneous bacterial peritonitis: An update. Mayo Clin Proc 1995;70:365–370.
71 Cooreman F: Pefloxacin vs cefuroxime for prophylaxis of infections after elective colorectal surgery (abstract). Rev Infect Dis 1989; 11(suppl 5):1301.
72 Gortz G, Boese-Landgraf J, Hopfenmuller W, Rodloff A, Kotwas J: Ciprofloxacin as single-dose antibiotic prophylaxis in colorectal surgery: Results of a randomized double-blind trial. Diagn Microbiol Infect Dis 1990:13:181–185.
73 Rohwedder R, Bonadea F, Benoti M, Ojea Quintana G: Single-dose oral ciprofloxacin plus parenteral metronidazole for perioperative antibiotic prophylaxis in colorectal surgery. Chemotherapy 1993;39: 218–224.
74 Keighley MR, Burdon DW: Identification of bacteria in the bile by duodenal aspiration. World J Surg 1978; 2:255–259.
75 Keighley MR: Micro-organisms in the bile: A preventable cause of sepsis after biliary surgery. Ann R Coll Surg Engl 1977;59:324–328.
76 Keighley MR, Drysdale RB, Quoraishi AH: Antibiotics in biliary disease: The relative importance of antibiotic concentrations in the bile and serum. Gut 1976;17:495–500.
77 Berquist EJ, Murphey SA: Prophylactic antibiotics for surgery. Med Clin North Am 1987;71:357–368.
78 Dellamonica P, Bernard E: Fluoroquinolones and surgical prophylaxis. Drugs 1993;45(suppl 3):102–113.
79 Cooreman F: Pefloxacin vs cefazolin as single dose prophylaxis in elective biliary tract surgery (abstract). Rev Infect Dis 1989;11(suppl 5): 1300.
80 Kujath P: Antibiotic prophylaxis in biliary tract surgery: Ciprofloxacin versus ceftriaxone. Am J Med 1989; 87(suppl 5A):255–257.
81 Anderson RE: Observations on the bacteriology of choledochal bile. Ann Surg 1951;133:456–489.
82 Eggert A: Die Bedeutung intraoperativer bakterieller Befunde bei Gallenblasen und Gallenwegsoperationen. Münch Med Wochenschr 1977; 119:955–958.
83 Blenkhorn JI: Streptococcal bacteremia in hepatobiliary operations. Surg Gynecol Obstet 1985;160:139–141.
84 Alveyn CG, Kilios G, Lowes JS, Robertson DAF: Oral antibiotic prophylaxis in high-risk patients undergoing ERCP. Gastroenterology 1991;100:308.

Chemotherapy

Chemotherapy 1966;42(suppl 1):54–61

Robert C. Moellering, Jr.

Department of Medicine,
Deaconess Hospital, and
Harvard Medical School,
Boston, Mass., USA

The Place of Quinolones in Everyday Clinical Practice

Key Words
Fluoroquinolones
Antimicrobial agents
Bacterial infections

Abstract
The new fluoroquinolones have many attributes and relatively few drawbacks. As a result, they have been rapidly adopted for the treatment of serious bacterial infections. The currently available fluoroquinolones are particularly effective for the treatment of urinary tract infections, bacterial diarrhea, enteric fever, gonorrhea, chancroid, skin and soft tissue infections, acute and chronic gram-negative osteomyelitis, gram-negative pneumonias, serious infections due to a variety of gram-negative bacilli, and in the eradication of the meningococcal carrier state. As we learn more about the properties of these compounds, the number and variety of infections for which they are useful continues to expand.

The new fluoroquinolones have rapidly attained the status of major therapeutic options against a broad variety of bacterial infections. The first of the 'quinolone' antimicrobial agents was discovered more than 30 years age, so this class of agents is not new [1]. Nalidixic acid and a number of the older quinolones including oxolinic acid, cinoxacin, pipemidic acid and others have been used in various parts of the world since the 1960s, but they have not achieved the status of first-line therapeutic agents. This is related to a number of drawbacks including a relatively narrow spectrum of activity which does not include *Pseudomonas aeruginosa, Serratia marcescens,* gram-positive cocci or anaerobes. A more significant drawback is that single-step mutations leading to high-level resistance to the older agents occur with frequencies in the range of 10^{-6} to 10^{-8} in a variety of bacterial species, and this has, on a number of occasions, led to the emergence of resistance during therapy with these drugs. A variety of adverse effects, related in particular to the gastrointestinal tract and central nervous system, are also seen relatively frequently with

KARGER

E-Mail karger@karger.ch
Fax + 41 61 306 12 34

0009–3157/96/0427–0054$10.00/0

Robert C. Moellering, Jr, MD
Department of Medicine
Deaconess Hospital
110 Francis Street, Suite 6A
Boston, MA 02215-5501 (USA)

the older quinolones. Not all of the characteristics of these older drugs are negative, however. They are generally well absorbed when given orally, are widely distributed among tissues and bodily fluids, and penetrate well into a variety of intracellular locations. They have relatively long serum half-lives and are bactericidal against a broad variety of microorganisms at therapeutic concentrations. Although minor side effects are common with this class of antimicrobials, they really do not have serious major-organ toxicity. Furthermore, despite more than 30 years of clinical use, there is still no confirmed report of microorganisms with plasmid-mediated or transferable resistance to these agents.

The 'modern quinolones' which have even more favorable properties are largely the results of chemical modifications of the older drugs based on ever-increasing knowledge of structure-activity relationships [2, 3]. The major changes which have led to the development of the new fluoroquinolones have been the addition of a fluorine atom at position 6 in the nucleus of the molecule and the addition of a piperazine or other ring structure at position 7. All of the currently available new fluoroquinolones contain these two structural features which, coupled with further modifications at position 1 and occasionally at positions 5 and/or 8, have led to the development of a series of compounds with markedly enhanced activity against gram-positive organisms and with excellent activity against *P. aeruginosa* and other gram-negative bacteria. Some of the new fluoroquinolones have good activity against anaerobes and such difficult-to-treat organisms as *Legionella* and mycobacteria, as well as 'atypical' respiratory pathogens including *Mycoplasma pneumoniae* and *Chlamydia pneumoniae*. Moreover, these organisms are 100- to 1,000-fold less likely to develop high-level resistance by single-step mutation to the newer fluoroquinolones than they were to the older agents. The new compounds are not subject to plasmid-mediated resistance and actually, under the appropriate circumstances, can cause loss of plasmids from bacteria in vitro and in vivo. They are bactericidal against many (but not all) pathogens and exhibit the oral bioavailability which is characteristic of this class of compounds. For a number of the currently available agents, parenteral preparations are also available, leading to greater flexibility in the design of therapeutic regimens. All of the new agents have long serum half-lives which allow once or twice daily dosing schedules, and they exhibit excellent penetration into tissues and phagocytic cells. They produce high concentrations in the urine and are well tolerated with a favorable side-effects profile. Compared to a variety of other new antimicrobials, the fluoroquinolones are relatively inexpensive and, since they can be given both orally and parenterally, are often cost-effective. The most important favorable attributes of these new compounds are listed in table 1. As shown in table 2, however, there are still

Table 1. Attributes of fluoroquinolones which make them attractive in the clinical setting

1	Broad spectrum of activity with particular potency against gram-negative bacilli including enteric pathogens.
2	Bactericidal against many (but not all) pathogens.
3	Low frequency of emergency of resistance during therapy (no transferrable mechanisms of resistance known).
4	Excellent oral bioavailability (with the possibility of parenteral administration for many compounds).
5	Long serum half-lives allow once or twice daily dosing.
6	Excellent penetration into tissues and phagocytic cells. High urine concentrations.
7	Well tolerated; favorable side-effects profile.
8	Relatively inexpensive: cost-effective.

Table 2. Disadvantages of fluoroquinolones *currently available* for clinical use

1	Relative lack of activity against gram-positive cocci (especially streptococci and enterococci), anaerobes, *Treponema pallidum.*
2	Emergence of resistance, especially among certain bacterial species (methicillin-resistant *Staphylococcus aureus* enterococci, *Campylobacter jejuni, Neisseria gonorrhoeae,* pseudomonads).
3	Adverse effects (drug/drug interactions, neurotoxicity, phototoxicity) with certain fluoroquinolones. Side effects often dose related.
4	Restrictions on use among children, pregnant women and nursing mothers.
5	Need to adjust dosage in persons with impaired renal function.

Table 3. Clinical areas in which fluoroquinolones are particularly effective

1	Urinary tract infections (including complicated infections and acute and chronic prostatitis).
2	Bacterial diarrhea and enteric fever.
3	Gonorrhea and chancroid.
4	Skin and soft tissue infections (including diabetic foot infections).
5	Acute and especially chronic (gram-negative) osteomyelitis.
6	Gram-negative pneumonias and bronchitis.
7	Gram-negative bacteremia (especially when due to multiresistant *Klebsiella* spp., *Enterobacter* spp., *Citrobacter* spp. and *Pseudomonas* spp.).
8	Other infections due to *Pseudomonas aeruginosa* (including malignant otitis externa); may require a second antimicrobial in some cases.
9	Meningococcal carriers.

disadvantages with the currently available agents. These include a relative lack of activity against gram-positive cocci (especially streptococci and enterococci) when compared with β-lactams and other drugs that specifically inhibit these organisms. In addition, the activity of the currently available compounds is less than optimal for anaerobes, and they lack useful clinical activity against *Treponema pallidum.* With the widespread clinical use of these compounds, we have begun to see the emergence of resistance (usually due to multiple-step mutations), especially among methicillin-resistant staphylococci, enterococci, *Campylobacter jejuni, Neisseria gonorrhoeae* and pseudomonads. Although adverse effects due to these compounds are not a major problem, some of the currently available agents do interfere with the metabolism of drugs such as theophylline, and mild gastrointestinal side effects, neurotoxicity and phototoxicity are seen with certain of the fluoroquinolones. A number of the latter effects appear to be dose related. Because these compounds do produce arthropathy of weight-bearing joints in certain experimental animals, their use in children and adolescents is still restricted. They are not approved for use in pregnancy or nursing mothers, although the teratogenic potential of these compounds appears to be low [4]. Finally, dosage adjustments are necessary for these compounds in patients with impaired renal function.

Table 3 lists a number of clinical areas in which the fluoroquinolones have been particularly effective to date. There is little question that they are important agents for the treatment of uncomplicated as well as complicated urinary tract infections. Clinical results obtained with the fluoroquinolones in head-to-head trials with older, established agents, such as the penicillins, cephalosporins, and trimethoprim-sulfamethoxazole, confirm that the fluoroquinolones are as effective or more effective than the older agents for the treatment of a variety of urinary tract infections. This is due, at least in part, to their achieving high and prolonged effective concentrations in the urine as well as tissues surrounding the urinary tract. Moreover, they are rapidly bactericidal against a variety of urinary pathogens (with the exception of enterococci). They have been shown to be effective for short-course (single-dose as well as 3-day) therapy of acute cystitis and uncomplicated urinary tract infections. Nonetheless, many experts currently recommend that these drugs should not be considered first-line therapy for uncomplicated cystitis, but should be reserved for patients with recurrent cystitis due to organsims resistant to the older agents and for patients with complicated urinary tract infections, including pyelonephritis. The fact that the fluoroquinolones penetrate well into prostatic tissue and have a spectrum of activity which includes the major pathogens involved in prostatic infections makes them particularly attractive for the treatment of acute and chronic prostatitis. Recently available data [5] confirm the effectiveness of the fluoroquinolones in acute bacterial prostatitis. Chronic prostatitis is much more difficult to treat and, until recently, it was unclear as to whether or not the fluoroquinolones had much to offer here. Nonetheless, recent data suggest that fluoroquinolones are effective in cases of chronic prostatitis due to *Escherichia coli* and other Enterobacteriaceae. Moreover, it appears that a duration of therapy as short as 1 month may be effective for chronic prostatitis due to Enterobacteriaceae. The role of the fluoroquinolones in treating chronic prostatitis due to more resistant organisms such as enterococci and *P. aeruginosa* remains to be determined.

Another area in which the fluoroquinolones have been particularly effective has been the treatment of bacterial diarrhea and enteric fever. Because (a) these drugs achieve high concentrations in the lumen of the gastrointestinal tract when given orally, (b) most of the currently available compounds do not significantly disturb anaerobic flora, and (c) their spectrum covers virtually all of the known bacterial causes of diarrhea, they are ideal agents for the treatment of bacterial enterocolitis. Indeed, except for furazolidone, the new fluoroquinolones are the only antibacterial agents which are effective against all known causes of bacterial diarrhea, although their activity against *Clostridium difficile* is marginal at best. They work well for traveler's diarrhea. Data show that they are also effective in treating *Salmonella* gastroenteritis and gastroenteritis due to *C. jejuni.* Unfortunately, however, there are now reports of the emergence of resistance to the fluoroquinolones among the latter organisms, and this may significantly impair their use in patients with suspected bacterial diarrhea of 'unknown' origin [6]. Data from a number of sources also suggest that the fluoroquinolones are quite effective for the treatment of enteric fever due to *Salmonella.* The role of fluoroquinolones in the treatment of *Salmonella* gastroenteritis remains controversial. There are data which suggest that these drugs are clinically effective in this setting, but the question of whether that rather marginal effectiveness is offset by the fact that they appear to prolong the carried state is a hotly debated issue. It is interesting that agents such as ciprofloxacin, norfloxacin and pefloxacin, which have been shown to be quite effective for the treatment of typhoid fever, are also effective in eradicating gastrointestinal carriage of *Salmonella typhi* if they are given for a sufficiently long course of treatment. Although the new fluoroquinolones have in vitro activity against *Helicobacter pylori,* they have not been effective when used alone for the eradication of *H. pylori* from patients with chronic gastritis, because of the rapid emergence of resistance in vivo.

The new fluoroquinolones have been shown to be highly effective in single-dose (and multiple-dose) treatment of uncomplicated gonococcal urethritis and cervicitis. The recent observation of the emergence of resistance to fluoroquinolones among gonococci in several parts of the world is disturbing [7]. Thus far, such isolates are comparatively rare, but if they become more widespread, they will significantly compromise the value of the fluoroquinolones for the treatment of gonococcal infections. Nonetheless, ciprofloxacin and ofloxacin have recently been recommended by the Centers for Disease Control in the United States as effective agents for the single-dose therapy of gonorrhea [8]. Those same recommendations, however, also suggest that coverage against possible coinfection with *Chlamydia trachomatis,* for example with doxycycline 100 mg orally twice daily × 7 days, be included in the regimen, because, although all of the presently available fluoroquinolones have in vitro activity against *C. trachomatis,* none of them has, to date, proven effective for single-dose therapy of this sexually transmitted disease. Among the currently available compounds, it appears that ofloxacin and fleroxacin are the most effective, but both must be given for a minimum of 7 days to be optimally effective. The fluoroquinolones are highly effective against *Haemophilus ducreyi* and, on the basis of current studies, it appears that they may well become the drugs of first choice for patients with chancroid because single-dose oral regimens are effective and there is no evidence for the emergence of resistance thus far. The fluoroquinolones have been less effective in the treatment of nonspecific urethritis and cervicitis and are not effective against *T. pallidum* in animal models. They are, therefore, not recommended for the treatment of syphilis and, in patients with sexually transmitted diseases at risk for coinfection with syphilis, alternative therapy should be given.

The broad spectrum of activity of the fluoroquinolones against a variety of pathogens which cause skin and skin structure infections and their oral bioavailability makes them particularly attractive for the treatment of such infections. They have been shown to be effective in the treatment of staphylococcal and streptococcal cellulitis, but the availability of other agents with good activity against these pathogens has led many clinicians to conclude that the fluoroquinolones are not drugs of first choice for uncomplicated cellulitis. However, they have been shown to be quite effective for the treatment of diabetic foot infections, either alone or in combination with an agent such as clindamycin or metronidazole that add activity against anaerobes [9]. These agents are obviously not a substitute for appropriate surgical management of these infections, but they are a very useful adjunct and have made it possible to treat a number of such infections on an outpatient basis. Another setting where the fluoroquinolones may be particularly useful is in the treatment of malignant otitis externa [10]. This infection, seen primarily in diabetic patients, can be life-threatening. A number of recent studies suggest that ciprofloxacin, in particular, may be useful in the management of this disease. For optimal effectiveness, the clinician must make absolutely certain that those patients with deep, invasive disease are identified and receive appropriate adjunctive surgical débridement when indicated. Moreover, the potential emergence of resistant organisms during therapy is always a worry when single-drug treatment is used for serious *P. aeruginosa* infections [10, 11]. In some studies, ciprofloxacin has been used in combination with rifampin for treatment of malignant external otitis [11]. Ciprofloxacin has also proven useful in the treatment of auricular perichondritis due to *P. aeruginosa* [12].

The fluoroquinolones have been effective in the therapy of acute and chronic osteomyelitis and septic arthritis [13]. These agents have been particularly attractive because of their spectrum of activity and because they can be given orally for long periods. The results thus far with fluoroquinolones for the treatment of acute and chronic gram-negative osteomyelitis have been, in general, favorable. Indeed, the presently available quinolones have become the mainstay of the treatment of osteomyelitis and infectious arthritis due to gram-negative bacilli. Acute osteomyelitis due to *Staphylococcus aureus* has also been treated successfully with a variety of fluoroquinolones, but emergence of resistance has been a problem, especially with methicillin-resistant staphylococci, and some investigators have suggested that rifampin should be added to a fluoroquinolone if these agents are to be used to treat osteomyelitis and/or septic arthritis due to staphylococci [13].

The outstanding intrinsic activity of the fluoroquinolones against a variety of Enterobacteriaceae, pseudomonads and other gram-negative bacilli has made them particularly attractive for the treatment of bronchitis and lower respiratory tract infections (particularly nosocomical pneumonias) due to gram-negative bacilli. They have also been used effectively in the treatment of gram-negative bacteremia, and they may be particularly useful in settings where multiresistant *Klebsiella, Enterobacter, Citrobacter,* or *Pseudomonas* spp. are suspected. These antimicrobials also possess excellent activity against *Neisseria meningitidis.* Since the fluoroquinolones produce high levels in nasal and nasopharyngeal secretions, they are potentially useful drugs for the management of meningococcal carriers. Both single-dose and short-course (2–5 days) therapy with ciprofloxacin has proven effective for the eradication of meningococci, and this drug may be particularly useful to treat nasopharyngeal carriers who harbor sulfonamide- or monocycline-resistant organisms [14, 15].

The role of the current fluoroquinolones in the treatment of community-acquired respiratory tract infections (otitis media, pharyngitis, sinusitis) and lower respiratory tract infections (including bronchitis and pneumonia) remains somewhat controversial. The major problem here relates to the fact that most of the currently available fluoroquinolones have relatively marginal activity against *Streptococcus pneumoniae.* There are clear-cut reports of the therapeutic success of these agents in treating respiratory tract infections due to pneumococci, but there are also a number of reports of failures. With this exception, the spectrum of activity of the fluoroquinolones, their ability to penetrate into respiratory secretions, and the other positive attributes noted earlier would make them ideal agents for the treatment of community-acquired respiratory tract infections. They provide excellent coverage against *Haemophilus influenzae* and *Moraxella catarrhalis,* and they possess activity against *M. pneumoniae, C. pneumoniae,* and *Legionella pneumophila.* Of the agents developed for clinical use thus far, temafloxacin had the greatest activity against the pneumococci and, before it was taken off the market because of toxicity, comparative studies of this agent showed it to be quite effective in the treatment of community-acquired pneumonia [16]. It is likely that the continued development and clinical deployment of newer fluoroquinolones with greater activity against gram-positive cocci, including *S. pneumoniae* and *Streptococcus pyogenes,* will solve this dilemma.

There are a number of other areas in which the fluoroquinolones have proven useful but are not drugs of choice. As noted earlier, they have activity against *Mycobacterium avium-intracellulare,* but attempts to use them as single agents in the treatment of infections due to these organisms in HIV-positive patients have resulted in the rapid emergence of resistant strains during therapy. A trial of intravenous ciprofloxacin for the treatment of febrile neutropenic episodes was terminated early by the EORTC because of failures in patients with gram-positive bacteremias [17]. A study with higher doses of ciprofloxacin suggested greater success, but the numbers in this study were too small to rule out β error [18]. Subsequent studies utilizing combinations of fluoroquinolones with β-lactams have overcome the problem of breakthrough bacteremias due to gram-positive organisms. There are limited examples of fluoroquinolone use for the treatment of bacterial endocarditis due to specifically susceptible organisms, bacterial meningitis, and ocular infections, and for the prophylaxis of infections in immunocompromised patients, but their exact role in these infections remains to be definitively determined. Likewise, the role of the presently available fluoroquinolones in the treatment of intra-abdominal and pelvic sepsis is problematic because they lack activity against anaerobes. The use of these agents in combination with antimicrobial agents possessing antianaerobic activity and the development of newer fluoroquinolones with better activity against anaerobes should obviate this difficulty in the future.

The development of the fluoroquinolones has resulted in a number of new antimicrobials that have rapidly become important therapeutic agents with many indications. Further refinements based on our understanding of structure-activity relationships are taking place and will continue over the coming decade. This will result in the development of fluoroquinolones with activity exceeding that of the agents currently available for clinical use. Nonetheless, it is very important that clinicians around the world do not overutilize these drugs simply because they are so easy to administer and cover so many different kinds of infections. Data already available suggest that overutilization will undoubtedly lead to the emergence of resistant organisms which, if not managed carefully, could destroy the clinical utility of this remarkable group of antimicrobial agents.

References

1 Lesher GY, Froehlich EJ, Gruett MD, Bailey JH, Brundage RP: 1,8-Naphthyridine derivatives: A new class of chemotherapeutic agents. J Med Pharm Chem 1962;5:1063–1065.

2 Chu DTW, Fernandes PB: Structure-activity relationships of the fluoroquinolones. Antimicrob Agents Chemother 1989;33:131–135.

3 Domagala JM: Structure-activity and structure-side-effect relationships for the quinolone antibacterials. J Antimicrob Chemother 1994;33:685–706.

4 Phillips I: Bacterial mutagenicity and the 4-quinolones. J Antimicrob Chemother 1987;20:771–773.

5 Naber KG: Role of quinolones in the treatment of chronic bacterial prostatitis; in Hooper DC, Wolfson JS (eds): Quinolone Antimicrobial Agents, ed 2. Washington, American Society for Microbiology, 1993, pp 285–297.

6 Rautelin H, Renkonen OV, Kosunen TU: Emergence of fluoroquinolone resistance in *Campylobacter jejuni* and *Campylobacter coli* in subjects from Finland. Antimicrob Agents Chemother 1991;35:2065–2069.

7 Knapp JS, Washington JA, Doyle LJ, Neal SW, Parekh MC, Rice RJ: Persistence of *Neisseria gonorrhoeae* strains with decreased susceptibilities to ciprofloxacin and ofloxacin in Cleveland, Ohio, from 1992 through 1993. Antimicrob Agents Chemother 1994;38:2194–2196.

8 Centers for Disease Control and Prevention: 1993 Sexually transmitted diseases treatment guidelines. MMWR Morb Mortal Wkly Rep 1993;42(No RR-14):57–62.

9 Peterson LR, Lissack LM, Canter K, Fasching CE, Clabots C, Gerding DL: Therapy of lower extremity infections with ciprofloxacin in patients with diabetes mellitus, peripheral vascular disease, or both. Am J Med 1989;86:801–808.

10 Lang R, Goshen S, Kitzes-Cohen R, Sadé J: Successful treatment of malignant external otitis with oral ciprofloxacin: Report of experience with 23 patients. J Infect Dis 1990; 161:537–540.
11 Rubin J, Stoehr A, Yu VL, Muder RR, Matador A, Kamerer DB: Efficacy of oral ciprofloxacin plus rifampin for treatment of malignant external otitis. Arch Otolaryngol Head Neck Surg 1989;115:1063–1069.
12 Noel SB, Scallar P, Meadors MC, Meek TJ, Pankey G: Treatment of *Pseudomonas aeruginosa* auricular perichondritis with oral ciprofloxacin. Dermatol Surg Oncol 1989;6: 633–637.
13 Lew D, Waldvogel F: Use of quinolones for treatment of osteomyelitis and septic arthritis; in Hooper DS, Wolfson JS (eds): Quinolone Antimicrobial Agents, ed 2. Washington, American Society for Microbiology, 1993, pp 371–379.
14 Dworzack DL, Sanders CC, Horowitz EA, Allais J, Sookpranee M, Sanders WE, Ferraro FM: Evaluation of single-dose ciprofloxacin in the eradication of *Neisseria meningitidis* for nasopharyngeal carrier. Antimicrob Agents Chemother 1988;32:1740–1741.
15 Renkonen OV, Sivonen A, Visakorpi R: Effect of ciprofloxacin on carrier rate of *Neisseria meningitidis* in army recruits in Finland. Antimicrob Agents Chemother 1987;31: 962–963.
16 Carbon C, Léophonte P, Petitpretz P, Chauvin JP, Hazebroucq J: Efficacy and safety of temafloxacin versus those of amoxicillin in hospitalized adults with community-acquired pneumonia. Antimicrob Agents Chemother 1992;36:833–839.
17 Meunier F, Zinner SH, Gaya H, Calandra T, Viscoli C, Klastersky J, Glauser M, and The European Organization for Research on Treatment of Cancer International Antimicrobial Therapy Cooperative Group: Prospective randomized evaluation of ciprofloxacin versus piperacillin plus amikacin for empiric antibiotic therapy of febrile granulocytopenic cancer patients with lymphomas and solid tumors. Antimicrob Agents Chemother 1991;35:873–878.
18 Johnson PRE, Liu Yin JA, Tooth JA: A randomized trial of high-dose ciprofloxacin versus azlocillin and netilmicin in the empirical therapy of febrile neutropenic patients. J Antimicrob Chemother 1992;30:203–214.

Subject Index

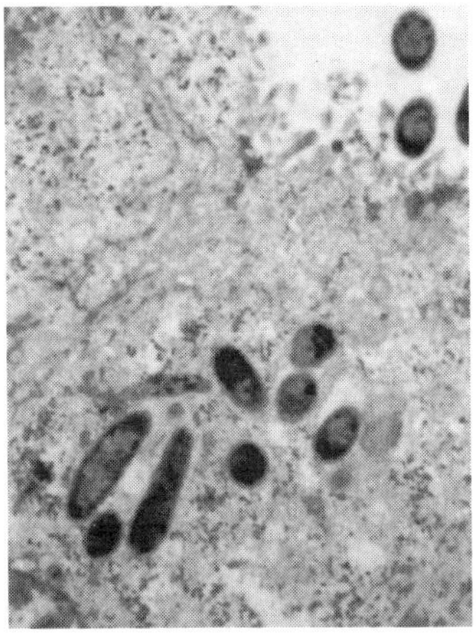

Comprehensive, compact and up-to-date

Yersiniosis: Present and Future

Editors: Giampietro Ravagnan, Claudio Chiesa

A comprehensive survey of current work in the field, this volume discusses recent changes in the clinical and epidemiological features of Yersinia infection in different countries, antimicrobial susceptibility of pathogenic Yersiniae isolated in various parts of the world and public health measures to prevent or control yersiniosis. Studies describe advances in the diagnosis of human and non-human infections using molecular sub-typing and DNA amplification, and new methods with culture. Sources of infection in the environment are discussed, as is research into therapeutic possibilities. The reader will also find the latest data on the genotypic characterization of the Yersinia species by PCR-based DNA fingerprinting, ribotyping and DNA-DNA hybridization.
Microbiologists and professionals in the areas of infectious disease and public health will appreciate the up-to-date information and expert commentary which this book provides.

Selected Contents

Epidemiology, Clinical Aspects and Public Health
Mollaret, H.H.: Fifteen Centuries of Yersiniosis • *Ostroff, St.:* Yersinia as an Emerging Infection: Epidemiologic Aspects of Yersiniosis • *Jensen, K.T. et al.: Yersinia enterocolitica Septicemia* in Denmark 1972–1991: A Report of 100 Cases. *Yersinia enterocolitica* Septicemia • *Suzuki, A. et al.:* Isolation of Yersinia from Wild Animals Living in Suburbs of Tokyo and Yokohama • *Aleksic, St. et al.:* Epidemiology of *Y. pseudotuberculosis* in Germany, 1983–1993 • *Fukushima, H.; Gomyoda, M.:* Putative Origin of *Yersinia enterocolitica* and *Yersinia pseudotuberculosis* in Japan • *Erickson, J.P.:* Fate of Yersinia

Advances in Diagnosis
Odinot, P.T. et al.: PCR-Based DNA Fingerprinting Discriminates between Different Biotypes of *Yersinia enterocolitica* • *Iteman, I. et al.:* High Genomic Polymorphism in *Yersinia pseudotuberculosis* • *De Giusti, M. et al.:* Occurrence of Pathogenic *Y. enterocolitica* in Slaughtered Pigs and Pork Products • *Kapperud, G.; Vardund, T.:* Detection of Pathogenic *Yersinia enterocolitica* in Food, Water, and Feces by Nested Polymerase Chain Reactions and Immunomagnetic Separation • *Dolina, M. et al.:* Molecular Typing of *Yersinia frederiksenii* Strains by Means of Ribotyping and DNA-DNA Hybridization • *Shiozawa, K. et al.:* Detection of *Yersinia pseudotuberculosis* DNAs in Paraffin-Embedded Tissues from Dead Chimpanzees by Using PCR • *Drysdale, E. M.; Bruce, J.:* Detection of Calcium Dependence in *Yersinia enterocolitica* by Conductance Microbiology

Therapy
Stolk-Engelaar, V. et al.: Activity of 24 Antimicrobials against *Yersinia enterocolitica* • *Preston, M. et al.:* Antimicrobial Susceptibility of Pathogenic *Yersinia enterocolitica* Isolated in Canada from 1972 to 1990

Immunology
Gripenberg-Lerche, Ch. et al.: Relation of YadA to Arthritogenicity of *Yersinia enterocolitica:* Experimental Studies in the Rat • *Easterbrook, T.J. et al.:* Studies on the Immunogenicity of the Pla Protein from *Yersinia pestis* • *Leary, S.E.C. et al.:* Expression of *Yersinia pestis* V Antigen in Attenuated *Salmonella typhimurium:* Development of a Novel Vaccine for Plague

Pathogenicity
Holmström, A. et al.: YopK, a Novel Virulence Determinant of *Yersinia pseudotuberculosis* • *Rakin, A.; Heesemann, J.:* Yersiniabactin/Pesticin Receptor: A Component of an Iron Uptake System of Highly Pathogenic Yersinia • *Fernández-Lago, L. et al.:* Immunological Study of the Expression of Proteins Encoded by the Virulence Plasmid of *Yersinia enterocolitica* Serotype O:9 • Hirota, *K.; Wake, A.:* Pathogenesis of *Pestis minor*

Genetic and Molecular Aspects
Ibrahim, A.: Genetic Diversity among *Yersinia enterocolitica* Strains as Revealed by Sequence Analysis of the 16S rRNA Gene • *Zhang, L. et al.:* Genetic Characterization of a Novel Locus of *Yersinia enterocolitica* Serotype O:8 for Down-Regulation of the Lipopolysaccharide O Side Chain at 37 °C • *Wren, B.W. et al.:* Genetic Analysis of phoP and ompR from *Yersinia enterocolitica, Yersinia pseudotuberculosis* and *Yersinia pestis*

Contributions to Microbiology and Immunology, Vol. 13
Series Editors: Cruse, J.M.; Lewis, R.E., Jr. (Jackson, Miss)
ISSN 0301–3081

Fields of Interest: Microbiology; Immunology; Infectious Diseases, Epidemiology, Genetics, Molecular Bilogy

Yersiniosis: Present and Future
6th International Symposium on Yersinia, Rome, September 1994
Editors: Ravagnan, G.; Chiesa, C. (Rome)
XII + 340 p., 75 fig., 92 tab., hard cover, 1995
CHF 344.– / DEM 412.– / USD 299.25
Prices subject to change
DEM price for Germany, USD price for USA only
ISBN 3–8055–6138–5

KI 95541

KARGER